Human Well-Being Values of Environmental Flows

Enhancing Social Equity in Integrated Water Resources Management

Cover: Hakaluki Haor, Northeast Bangladesh (photo by Karen Meijer)

Human Well-Being Values of Environmental Flows

Enhancing Social Equity in Integrated Water Resources Management

Karen S. Meijer

Delft Hydraulics Select Series 10/2007

This research has been supported by the Delft Cluster programme within the Dutch ICES funding. The research has been carried out within the framework of the Netherlands Centre for River Studies (NCR).

© 2007 K.S. Meijer and IOS Press

All rights reserved. No part of this book may be reproduced, stored in a retrieval system, or transmitted, in any form or by any means, without prior permission of the publisher.

ISBN: 978-1-58603-732-1

Keywords: environmental flows, human well-being, integrated water resources management, social equity.

Published and distributed by IOS Press under the imprint Delft University Press

Publisher
IOS Press
Nieuwe Hemweg 6b
1013 BG Amsterdam
The Netherlands
Tel: +31 20 688 3355
Fax: +31 20 687 0019
Email: info@iospress.nl
www.iospress.nl
www.dupress.nl

LEGAL NOTICE
The publisher is not responsible for the use which might be made of the following information.

PRINTED IN THE NETHERLANDS

Summary

Introduction

For centuries, people have altered the natural flow of rivers, for example through the construction of reservoirs, to improve the well-being of people. During the last century, negative impacts on downstream ecosystems as a result of upstream water resources development interventions in the river system became apparent. Because river ecosystems are a source of income, food and other goods and services for millions of people all over the world, many local communities have suffered as a result of these water resources developments. This means that although water resources development has undoubtedly brought prosperity to a large number of people, these benefits were often not equally shared among the people in the river basin.

Environmental flows can be defined as the proportion of the natural flow regime that is maintained in a river, wetland or coastal zone to sustain ecosystems and the benefits they provide for people. Various methods have been developed to quantify the environmental flow requirement. Most of these environmental flow assessment methods focus on the relationship between river flows and ecosystem condition, and do not explicitly quantify the importance of flows for the well-being of various groups of people.

To balance water allocation in Integrated Water Resources Management (IWRM), it is necessary to quantify all benefits and negative impacts of allocating water to a certain use sector. Traditionally, the focus was on economic benefits, but nowadays also environmental sustainability and social equity are important criteria for evaluating water resources management strategies. In order to be able to assess social equity, defined as the distribution of positive and negative impacts over various stakeholder groups, also the importance of environmental flows for the well-being of each of these stakeholder groups must be quantified.

Methods to quantify the value environmental flows have for human well-being (further referred to as human well-being values of environmental flows) are currently not available, yet are urgently required to facilitate and enhance equitable water resources management.

The research objective for this thesis was therefore:

> *To develop an approach for the assessment of human well-being values and social equity related to environmental flows for application in Integrated Water Resources Management.*

To achieve this objective, a conceptual model was developed which describes the relationship between human well-being, the river ecosystem and river flows. This conceptual model was then applied in two case studies, the first on the Surma-Kushiyara river floodplains in Bangladesh, and the second on the Hamoun wetlands in Iran. Both case studies served to test and further refine the conceptual model.

Conceptual model

To describe *human well-being*, various components of this concept were distinguished. Three components are directly related to the river ecosystem: 1) income & food, 2) health and 3) perception and experience. Because the socio-economic system is complex and dynamic, changes in these three components of well-being are likely to impact other components of well-being such as independence, social structure and other psycho-social factors. These components of well-being are in this thesis referred to as second order values, because they are only indirectly affected by the changes in the river ecosystem. The components directly related to the ecosystem are referred to as first order. As a result of changes in the second order components, the first order components can change further, either in a positive or in a negative direction. The quantification in this thesis focused on the first order values.

Quantification of impacts of flow regime changes on human well-being requires a thorough conceptual understanding of the *relationships* between river flows, the river ecosystem and human well-being. Although the focus of this thesis is on the relationship between human well-being and the river ecosystem, it was considered necessary to include also the relationship between the river ecosystem and the river flow regime in the conceptual model. Both relationships are necessary to assess changes in human well-being resulting from changes in the flow regime. Moreover, only through assessing all relationships problems in connecting different concepts and approaches could be identified.

In order to assess the importance of the river flow regime and the river ecosystem for the well-being of various groups of people, it is necessary to assess not only the relationship between human well-being, the river ecosystem and the river flow regime, but also the *context* of these relationships. The context consists of all factors that can influence these relationships. An example of the context is access to ecosystem goods and services. When poor fishermen do not have access to certain fishing grounds, maintaining this part of the ecosystem will not benefit these particular fishermen. Only through giving ample consideration to the context, the importance of the river ecosystem, and of river flows, for human well-being can be understood.

Both the relationships and the context are included in the conceptual model that was developed in this thesis (see Figure S1).

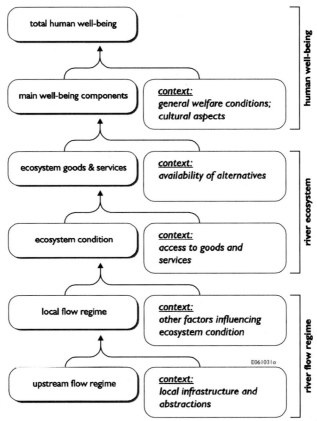

Figure S.1 The conceptual model including the relationships between human well-being, the river ecosystem and the river flow regime, and the context of these relationships.

The purpose of the conceptual model is to support the assessment of impacts on human well-being values and of the social equity resulting from alternative water resources management strategies. It is important to establish the links of the conceptual model for the various groups of people that are likely to experience different effects from changes in the river flow regime. Identifying these groups of people (the stakeholder groups) is therefore an important first step in the application of the conceptual model.

Applying the conceptual model involves a total of five steps:

1. identifying stakeholder groups;
2. assessing the relationship between human well-being and the river ecosystem;
3. assessing the relationship between the river ecosystem and the local flow regime;
4. assessing the relationship between the local flow regime and the upstream flow regime; and
5. estimating impacts of water resources management measures on the well-being of the identified stakeholder groups.

Case studies

Surma and Kushiyara Rivers and floodplains, Bangladesh

In the Surma and Kushiyara Rivers in Northeast Bangladesh the flow regime is expected to change as a result of three developments:

- Morphological changes at the location where the Barak River enters Bangladesh from India and bifurcates in the Surma River and the Kushiyara River.
- The construction of the Tipaimukh reservoir in the Barak River (upstream of the India-Bangladesh border), assumed to result in reduced peak flows and increased low flows.
- The diversion of water for irrigation of the Cachar Plain from the Barak River (also upstream of the India-Bangladesh border). In combination with the Tipaimukh reservoir, this development is assumed to result in a further reduction of peak flows. The increase in low flows resulting from the Tipaimukh dam will be limited. It is not clear whether the net effect on the low flows will be an increase or a decrease of flows.

The conceptual model has been applied to assess the importance of the river flow regime for the approximately 300,000 people living in the Surma-Kushiyara floodplain. Data collection for the link between human well-being and the ecosystem was done through interviews at the household level. For the other links existing data were used.

Three main types of stakeholders can be distinguished in the Surma-Kushiyara floodplain: farmers, fishermen and others. Farmers and fishermen depend to a large extent on the goods and services of the river and floodplain for their income and food. Approximately 10% of the inhabitants uses the river water for domestic purposes. These people are generally the poorest: the fishermen and the landless labourers. The main link between the river ecosystem and perception and experience consists of inconveniences related to flooding of the settlement area. Also for this well-being value, the poorest people are the most affected.

In a qualitative analysis the three types of changes to the flow regime mentioned in the above were considered. The analysis reveals that especially fishermen will loose income when the floodplains are not flooded in time. On the other hand, the fishermen and other poor households' health will benefit from an increase of river flows during the dry season.

Hamoun wetlands, Iran

A second application of the conceptual model was carried out as part of an IWRM study in the Sistan delta in Iran. This delta is bordered by the Hamoun wetlands system, with an area of 500,000 ha under wet conditions. The wetlands are of great value for both the people who use the wetland for income, as well as for other people who profit from the prevention of sandstorms and from the scenery for celebrating, amongst others, their New Year festivities.

The five steps of the conceptual model were followed to make a quantitative assessment of the well-being values under various water resources management strategies. Group discussions and a questionnaire survey were conducted to collect data on the link between human well-being and the wetland ecosystem. For the other links, the results of other components of the IWRM study were used.

With regard to their dependence on the Hamoun wetlands, the population of the Sistan delta can be divided into three groups. Of the total of approximately 71,000 households, around 15,000 households directly depend for more than 70% of their income on the Hamoun wetlands, around 30,000 households depend for more than 70% of their income on irrigated agriculture, while the income of the remaining 26,000 urban households is largely not directly related to water.

The main ecosystem goods and services important for the population are the availability of fish, birds and reeds to support income and food production, as well as the prevention of sandstorms and the regulation of the local climate. Five important hydrological parameters were identified to sustain these goods and services. For the sustenance of fish, reeds and birds these are wetland spills, regular droughts, a minimum inundated area in fall, and a minimum volume of water inflow in spring. It was considered important to have a minimum area of Hamoun-e-Saberi inundated from May till August to prevent sandstorms.

Water resources management strategies can affect the well-being of the identified stakeholder groups through their impacts on the wetland hydrological regime, subsequent changes in the ecosystem, and resulting availability of ecosystem goods and services. This thesis discusses an approach to quantify the change in each well-being component. With the resulting relationships the impacts of alternative water resources management strategies on the diverse stakeholder groups and the equity in impact distribution was assessed.

The main water demands come from irrigated agriculture and the Hamoun wetlands. Both water use sectors are important for the economy of the region and the well-being of the stakeholder groups. A balance needs to be found in the distribution of water over these two demands. The strategies analysed focussed on increasing and decreasing the irrigated area to investigate this balance. It was found that in the present situation the distribution of impacts is most equitable. Although in the present situation and in the situation with a decreased irrigated area the ecosystem condition was calculated to be the same, the years with reduced income for the various stakeholder groups varied. This indicates that assessing human well-being impacts provides useful additional information which enhances informed and equitable management.

Conclusions
Through describing both the relationships between human well-being, the river ecosystem and the river flow regime, and the context, the proposed conceptual model will facilitate the quantification of changes in the river flow regime or the river ecosystem in terms of human well-being and social equity as part of an IWRM analysis.

While the conceptual model and stepwise approach are generally applicable, the data collection and analysis methods need to be tailored to the situation. In arid Sistan, it was easy to draw the border of the wetland ecosystem and identify the people who use the wetland. In the wet Northeast of Bangladesh, it proved more difficult to identify to what extent ecosystem goods and services depended on river flows, and therefore it was more difficult to understand whether the users of these goods and services were stakeholders with regard to the flow regime. Moreover, the fact that in Sistan the interventions to the river system had taken place already, allowed for a direct discussion with the stakeholders of the various impacts of the changes to the wetland for their well-being. In the Surma-Kushiyara floodplain, where changes had not taken place yet, data collection focussed on current use of the ecosystem and understanding the context in order to predict well-being impacts of flow regime changes.

In the case studies it was found that through assessing the human well-being values of the river ecosystem, additional useful information was generated to aid decision-making. It is therefore recommended that such an assessment is conducted as part of any IWRM study. When limited time and resources are available the human well-being assessment can be carried out as a qualitative assessment only, focussing on the identification of stakeholder groups and the importance of various ecosystem goods and services for each of these groups. If time and resources are sufficiently available, the quantification of first order impacts as described in this thesis can be extended with a quantification of second order impacts.

The overall conclusion is that this thesis provides an approach for assessing human well-being values of environmental flows as part of IWRM studies, consisting of a conceptual model together with a stepwise approach for the actual assessment. With this approach the thesis makes a contribution to the further operationalisation of the concepts of IWRM and environmental flows. The approach presented in this thesis generates comprehensive and socially-relevant information to decision-makers, which is essential to enhance social equity in IWRM.

Contents

SUMMARY V
CONTENTS XI

1 INTRODUCTION: THE NEED TO INCLUDE HUMAN WELL-BEING VALUES OF RIVER ECOSYSTEMS IN INTEGRATED WATER RESOURCES MANAGEMENT 1

1.1 Limits to water resources development 1
1.2 The importance of natural ecosystems for people 1
1.3 Ecosystem water needs in Integrated Water Resources Management 2
1.4 Challenge for this thesis and research objective 4
1.5 Scope 5
1.6 Research questions and research approach 7

2 HUMAN WELL-BEING IN CURRENT APPROACHES IN THE FIELD OF ENVIRONMENTAL FLOWS AND INTEGRATED WATER RESOURCES MANAGEMENT 9

2.1 Introduction 9
2.2 Human well-being in environmental flows approaches 9
2.3 Human well-being in Integrated Water Resources Management 17
2.4 Environmental flows in IWRM 24
2.5 Conclusions: Challenge for this thesis 27

3 A CONCEPTUAL MODEL FOR THE RELATIONSHIP BETWEEN HUMAN WELL-BEING AND ENVIRONMENTAL FLOWS 29

3.1 Introduction 29
3.2 Development of a conceptual model linking human well-being, the river ecosystem and the river flow regime 29
3.3 Human well-being 30
3.4 The river ecosystem and the link with human well-being 36
3.5 The river flow regime and the link with the river ecosystem 40
3.6 The context: understanding the importance of river flows and related ecosystems for people 43
3.7 Applying the conceptual model as part of IWRM analysis: a stepwise approach ... 47
3.8 Conclusions 55

4 WELL-BEING VALUES OF THE SURMA AND KUSHIYARA RIVERS AND FLOODPLAIN, BANGLADESH 57

4.1 Introduction 57
4.2 Background & objective of the case study 58
4.3 Method: approach & data collection 61
4.4 Identifying stakeholder groups 65
4.5 Relationship between human well-being and the river and floodplain ecosystem 67
4.6 Relationship between the river and floodplain ecosystem and the local flow regime 82
4.7 Relationship between the local flow regime and the upstream flow regime 89
4.8 Combining all: impacts of flow regime changes on human well-being 91
4.9 Discussion and conclusions 97

5 WELL-BEING IMPACTS OF CHANGES IN HYDROLOGY OF THE HAMOUN WETLANDS, IRAN 101

- 5.1 INTRODUCTION 101
- 5.2 BACKGROUND AND OBJECTIVE OF THE CASE STUDY 101
- 5.3 METHOD: APPROACH & DATA COLLECTION 104
- 5.4 IDENTIFYING STAKEHOLDER GROUPS 109
- 5.5 RELATIONSHIP BETWEEN HUMAN WELL-BEING AND THE HAMOUN WETLANDS ECOSYSTEM ... 115
- 5.6 RELATIONSHIP BETWEEN HAMOUN WETLANDS ECOSYSTEM AND THE HAMOUN HYDROLOGY 126
- 5.7 RELATIONSHIP BETWEEN THE HAMOUN WETLAND HYDROLOGY AND THE UPSTREAM FLOW REGIME 132
- 5.8 COMBINING ALL: IMPACTS OF CHANGED HAMOUN HYDROLOGY ON THE WELL-BEING OF THE IDENTIFIED STAKEHOLDER GROUPS 136
- 5.9 DISCUSSION & CONCLUSIONS 139

6 PRACTICAL APPLICATION OF THE CONCEPTUAL MODEL: LESSONS FROM THE CASE STUDIES 143

- 6.1 INTRODUCTION 143
- 6.2 STEP 1: IDENTIFYING STAKEHOLDER GROUPS 144
- 6.3 STEP 2: ASSESSING THE RELATIONSHIP HUMAN WELL-BEING AND THE RIVER ECOSYSTEM 145
- 6.4 STEP 3: ASSESSING THE RELATIONSHIP BETWEEN THE RIVER ECOSYSTEM AND THE LOCAL FLOW REGIME 150
- 6.5 STEP 4: ASSESSING THE RELATIONSHIP BETWEEN THE LOCAL FLOW REGIME AND THE UPSTREAM FLOW REGIME 151
- 6.6 STEP 5: ESTIMATING IMPACTS OF WATER RESOURCES MANAGEMENT MEASURES ON THE WELL-BEING OF THE IDENTIFIED STAKEHOLDER GROUPS 152
- 6.7 SUGGESTED ADDITIONAL STEPS 152
- 6.8 CONCLUSIONS 153

7 DISCUSSION AND CONCLUSIONS 155

- 7.1 INTRODUCTION 155
- 7.2 RESEARCH QUESTION 1: CURRENT APPROACHES 155
- 7.3 RESEARCH QUESTION 2: CONCEPTUAL RELATIONSHIP AND LINK WITH ENVIRONMENTAL FLOW ASSESSMENTS AND INTEGRATED WATER RESOURCES MANAGEMENT 156
- 7.4 RESEARCH QUESTION 3: QUANTIFICATION IN PRACTICAL APPLICATIONS 157
- 7.5 CONTRIBUTION TO SCIENCE AND SOCIETY 158
- 7.6 FINAL CONCLUSION 161
- 7.7 FUTURE RESEARCH 161

REFERENCES 163

SAMENVATTING 173

BENGALI SUMMARY 179

FARSI SUMMARY 185

ACKNOWLEDGEMENTS 193

CURRICULUM VITAE 195

1 Introduction: the need to include human well-being values of river ecosystems in Integrated Water Resources Management

1.1 Limits to water resources development

Water is of vital importance for human life. To improve the availability of water for society, water resources have been stored and diverted for centuries. Already in the 12th century AD the Sri Lankan king Parakrama Bahu I expressed the importance of water for human well-being and the need for water resources development:

> *"Let not a single drop of water that falls on the land go into the sea without serving the people."*

This desire to alter river ecosystems for the benefit of people was still the general view on water resources development in the early 20th century, according to Churchill's wish for the Nile (McCully, 2001):

> *"One day, every last drop of water which drains into the whole valley of the Nile ... shall be equally and amicably divided among the river people, and the Nile itself ... shall perish gloriously and never reach the sea."*

There is little doubt that water resources development has indeed improved the well-being of numerous people (Barker et al., 2000; Rijsberman, 2003). However, around the 1950s the other side of water resources development became apparent: river ecosystems degraded with as obvious loss the disappearance of game fish (McCully, 2001; Postel & Richter, 2003). A strong example of the result of over-exploiting of rivers is the desiccation of the Aral sea. The Aral sea borders have within 50 years retreated over 100 km from its original boundary, with reduced income from fisheries, increased sandstorms and reduced human health as a result (Glantz & Zonn, 2005). The degradation of natural ecosystems revealed that river flows cannot infinitely be withdrawn and modified, and that in fact water resources development has its limits. This poses water resources managers for some difficult questions: what are these limits? How important are these river ecosystems and how does their value weigh out against the benefits of development?

1.2 The importance of natural ecosystems for people

Natural ecosystems provide various goods and services essential to people (IUCN, 2000; Millennium Ecosystem Assessment, 2005). Various studies have attributed considerable economic values to the goods and services provided by ecosystems in general (e.g. Costanza et al., 1997), and wetlands in particular (e.g. Schuyt & Brander, 2004). Abstracting and modifying ecosystems for development purposes can only be sustainable when the ecosystem itself is maintained at a certain level. The importance of maintaining ecosystems was officially recognised in 1987, in the definition of sustainable development: sustainable development is development that meets the needs of the present without compromising the ability of future generations to meet their own needs (WCED, 1987). Although development creates livelihoods for poor communities, it is at the same time the unmodified ecosystem on which often the poorest people of the world depend (Silvius et al., 2000; DFID, 2001; Mainka et al., 2005).

In Africa, various communities depend on the rising and receding floods for the cycle of fisheries, recession agriculture, and livestock herding (Drijver & Marchand, 1986; Marchand, 1987; Fiselier, 1990). This is for example the case in the Hadejia-Nguru wetlands of Nigeria (Thompson & Polet, 2000), the Logone floodplain in Cameroon (Mouafo et al., 2002), and the Inland Delta of the Niger in Mali (Marchand, 1987). Also in Asia, for example along the Mekong, these dependencies of poor communities on river ecosystems are prevalent (MRC, 2003).

Upstream diversions and regulations of river flows for the purpose of development, have in many situations led to degradation of the downstream river ecosystems, including wetlands and floodplains, with subsequent impoverishment of the communities depending on the resources provided by these ecosystems (Horowitz, 1991). The Manantali dam of Senegal reduced floods leading to a decreased production in recession agriculture (Adams, 1999). As a result of the construction of the Maga dam and irrigation infrastructure in the 1980s, flooding of the Logone floodplain in Cameroon decreased, leading to reduced yields from fishing, reduced grazing capacity and a changed wildlife habitat. Many floodplain-pasture dependent cattle rearers were forced to leave the area (Mouafo et al., 2002). A program for flood restoration has successfully brought back some of the goods and services (Loth, 2004). The World Commission on Dams (2000) states that "Millions of people living downstream from dams – particularly those reliant on natural floodplain function and fisheries –have also suffered serious harm to their livelihoods and the future productivity of their resources has been put at risk". As possible reasons for the lack of attention for impacts on downstream communities, Adams (2000) identifies the fact that these people live spread over a large area, do not form one well-organised group, live far downstream of diversion structures, or have requirements that are diffuse, intangible or difficult to quantify.

1.3 Ecosystem water needs in Integrated Water Resources Management

1.3.1 Balancing water for ecosystems against other users

Integrated Water Resources Management (IWRM) deals with the balancing of various interests with regard to the water resources system. With a growing world population and increased standard-of-living, water becomes more and more scarce, with as a result an increased pressure on ecosystems (Cosgrove & Rijsberman, 2000; United Nations, 2003). Logically, allocating more water to nature means less water for other users. Irrigated agriculture accounts for 65%-75% of the total amount of water withdrawn for human use (Wallace et al., 2003). The water balance discussion therefore often focuses on the distribution of water between irrigated agriculture and natural ecosystems (Rijsberman & Mohammed, 2003). To make an informed trade-off between the various water demands it is necessary to understand the water demands of river ecosystems as well as the value river ecosystems present for society.

1.3.2 Water demand of river ecosystems: environmental flows

To discuss impacts of water resources development on river ecosystems two concepts are useful: the river continuum concept and the flood pulse concept. In the *river continuum concept* the focus is on the connection between upstream, middle and lower reaches of the river (Vannote et al., 1980). Next to being of utmost importance for migrating fish, this continuum also ensures transport of nutrients and sediments which create living opportunities for downstream organisms. The *flood pulse concept* stresses the importance of various components of the natural flow regime (Junk et al., 1989). Alternate high and low flows cause the river to make a lateral movement. Through the year the width of a river changes and levees and floodplains become inundated. As a result of the inundation, various biological processes are activated and habitat opportunities for various species are created. Fish can migrate between the river channel and the floodplain. Both concepts are relevant to maintain river-dependent ecosystems (Bunn & Arthington, 2002). However, to answer the question of how much water should be available downstream of interventions or abstractions, the flood pulse concept is the most suitable concept. In this concept, natural ecosystems can be considered to be the result of the prevailing flow regime. The flow regime required to maintain an ecosystem in a desired condition is commonly referred to as the environmental flow requirement (EFR) or simply as environmental flows (EF). Various methods have been developed to assess these flow regimes, which will be discussed in more detail in Chapter 2.

1.3.3 Value of river ecosystems

Most of the studies on environmental flows have focused on the relationship between river flows and the natural ecosystem. Until recently, not much attention has been paid to the importance of river ecosystems for local communities (Pollard & Simanowitz, 1997; Pollard, 2002; Scudder, 2002). Nyambe and Breen (2002) point out that although in many cases there will be an implicit assumption that improving the river ecosystem will automatically sustain livelihoods, due to the dynamic nature of social systems, assessing environmental sustainability on its own cannot guarantee livelihood sustainability. They argue that an explicit assessment of human impacts of changed river ecosystem dynamics is required.

Economic valuation of ecosystems is a topic that has received ample attention over the past decades. An often applied framework for assessing ecosystem values is the 'total economic value' framework which distinguishes three types of use values (direct, indirect and optional), as well as non-use values (Barbier et al., 1997). For several ecosystems, economic values have been assessed to show the economic losses that would take place in case the ecosystems are degraded (e.g. Gram et al., 2001; Schuyt, 2005). Through economic valuation, Barbier and Thompson (1998) have contended that gains estimated from an irrigation project upstream from the Hadejia-Jama'are floodplain in Nigeria could not outweigh the losses in direct use value of the wetland products. A similar comparison was made between food production in the floodplain wetlands of the Inland Niger delta in Mali and an upstream irrigation area (Marchand, 1987).

Ecosystem valuation provides crucial information for decision-making (Emerton & Bos, 2004). However, the concept has some drawbacks as well. Although the total economic value framework claims to assess the total economic value, in practice only partial values are assessed. In developing countries the focus seems to be on the tangible, marketable values (see e.g. Klasen, 2002), whereas in developed countries studies focus on the non-use values, mainly of recreation (see e.g. Douglas & Johnson, 1991; Loomis, 1998; Hickey & Diaz, 1999; Whittaker & Shelby, 2002). Another major drawback is the focus on monetary values. With regard to impacts on dam-displaced communities, Cernea (2000) argues that cost-benefit analysis may lead to decisions based on "the greater good for the greater numbers", which conflicts with social justice, and as a result of which the poor may loose out (GWP, 2003). Moreover, if in a certain case subsistence use of wetlands has a low value in economic terms, this may still mean one hundred percent of a household's food and income. Therefore, a completely economic comparison cannot always be morally justified. Besides the tangible products local communities use for their food and income, river ecosystems are likely to contribute to other aspects of peoples' life such as good health and a pleasant environment. It is necessary to recognize that ecosystem goods and services are important in different ways to different groups (Wallace et al., 2003). Assessing ecosystem values asks for a wider range of criteria than only monetary ones.

Also in literature on Integrated Water Resources Management, it is recognised that economic efficiency alone is not sufficient to base decisions on. The current vision is that IWRM should also contribute to environmental sustainability and social equity (GWP, 2000; ESCAP, 2000). Social equity requires a consideration of how impacts are distributed over various groups of people. Since the formulation of the Millennium Development Goals, various aspects of poverty are high on the agenda. To make water resources management more equitable, it is argued that the contribution of water to livelihoods and to the alleviation of poverty should receive explicit attention in IWRM (Soussan, 2004a; Soussan, 2004b).

1.4 Challenge for this thesis and research objective

Water resources development has undoubtedly brought prosperity for a large number of people. However, these benefits were often not equally shared among the people in the river basin: local communities have suffered from the negative impacts of these developments on the river ecosystem that they depend upon.

Environmental flow assessment methods have over the last decades emerged as a means to quantify the water need for ecosystems. IWRM has emerged over the same period to deal with the interest of various stakeholders in balancing water supply and demand. An assessment of the impacts on downstream communities resulting from development-induced ecosystem changes will contribute to a better representation of the value of ecosystems and enhance social equity in IWRM. To be able to consider the link between river flows, ecosystems and impacts on people to take equitable decisions in water resources management, it is required that environmental flow assessment methods are extended to explicitly address social aspects. Such methods to quantify the well-being value in relation to different flow regimes are currently not available, yet are strongly required to facilitate and enhance equitable water resources management.

The objective for this thesis is therefore formulated as:

> *To develop an approach for the assessment of human well-being values and social equity related to environmental flows for application in Integrated Water Resources Management.*

Through providing an approach for assessing the human well-being value of environmental flows, this thesis hopes to contribute to informed and equitable water resources management, and in this way to contribute to the protection of both the river ecosystem and the well-being of the people depending on this ecosystem.

1.5 Scope

1.5.1 Focus on benefits of river flows and river ecosystems

The relationships between people and ecosystems can take various forms (Figure 1.1): people can influence ecosystems in a positive way, through environmental management, and in a negative way, as a result of interventions and pollution. Ecosystems, in turn, provide various goods and services for people, but also pose threats, like floods and droughts. In this thesis only one of these relationships is central: the environmental goods and services provided to people by river ecosystems.

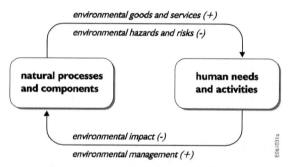

Figure 1.1 Relationships between men and ecosystems (Source: De Groot, 1992a)

1.5.2 Focus on impacts of changes in downstream ecosystems

Flow regime changes will have impacts in various locations and on various groups of people. The changes in the flow regime normally result from measures to improve human well-being, for example the construction of a reservoir for irrigation or hydropower generation. In this thesis the focus is on people who make use of the natural river ecosystem downstream of the interventions, and not on people that directly benefit from the interventions.

1.5.3 Focus on local benefits in developing countries

Nature benefits people at various spatial scales: subsistence of local communities, regional provision of products for the market, or a global appreciation of biodiversity conservation. Nature is a prerequisite for human beings to exist at all and therefore could be considered at a global scale. However, local communities often have a more direct relationship with the river ecosystem, and may be more severely impacted when the river ecosystem changes. These local communities are the focus of this thesis. Such strong relationships with the natural ecosystem for human well-being are mainly found among poor communities in developing countries. As a result the focus lies on developing countries.

1.5.4 Focus on instream uses and small scale abstractions for direct human needs

The focus of this thesis is on the instream benefits of river flows through natural ecosystems. However, the focus is also on local communities. For local communities, some instream benefits of river flows have no direct relationship with the quality of ecosystems. They include water use for boating, washing, and small scale abstractions for domestic use. Although these types of river water use are not actually part of environmental flows, to generate a complete picture of the importance of river flows for the well-being of local communities, it was decided to include such non-ecosystem dependent uses in the analysis.

1.5.5 Focus on water quantity

Although water quality is of high importance for both ecosystems and human well-being, the focus in this thesis is on water allocation issues, and hence on water quantity. A complication is the fact that water quantity plays a role in water quality as well, through transporting waste, diluting pollution loads and re-aerating the water body. This function of the river flow regime will be mentioned where relevant, but not investigated in detail.

1.5.6 Focus on changes

The aim in this thesis is not to assign a total well-being value to a river ecosystem, but rather to assess changes in well-being as a result of changes in the river ecosystem induced by altered river flow regimes.

1.5.7 Focus on impacts of interventions as planned

Various negative impacts of development projects on human communities are due to non-compliance with operation rules or compensation agreements. Taking into account the risk of non-compliance in an early stage may lead to the decision not to carry out a project. Despite the high relevance of such impacts, these are not considered in this thesis. The focus is on the impacts of projects according to plan, in order to support decisions on operation rules, compensation, mitigation or benefit-sharing.

1.6 Research questions and research approach

To achieve the research objective mentioned in Section 1.4, the following questions need to be answered:

1. What are the approaches currently used in environmental flow assessments and Integrated Water Resources Management, and how is or can human well-being be linked to these approaches?
2. What is the conceptual relationship between river flows, the river ecosystem and human well-being and social equity, and how does this conceptual relationship fit in the approaches in environmental flow assessments and Integrated Water Resources Management?
3. How can the conceptual relationships be quantified in practical applications?

To answer the *first research question*, Chapter 2 discusses the current approaches in environmental flow assessments and in Integrated Water Resources Management, based on literature. In addition, it discusses how human well-being is addressed in these approaches. The result of Chapter 2 is a description of what is currently missing to assess human well-being related to environmental flows in Integrated Water Resources Management, together with an understanding of what aspects a method to assess human well-being related to environmental flows should comply with. This forms the starting point for the theory developed in Chapter 3, to answer the *second research question*. The proposed conceptual model is developed based on literature and on the experiences in two case studies. These case studies, the first on the Surma and Kushiyara Rivers in Northeast Bangladesh, and the second on the Hamoun wetlands in Iran, are discussed in Chapters 4 and 5 respectively. In addition to testing of the conceptual model, the case studies provided insight in suitable methods for practical application of the model. The lessons learned through these case studies are translated into guidelines for future assessments in Chapter 6 to answer the *third research question*. Chapter 7 discusses the applicability of the developed theory, the answering of the research questions, and the contribution this study can make to water resources management and to the protection of river ecosystems and associated human well-being.

2 Human well-being in current approaches in the field of environmental flows and Integrated Water Resources Management

2.1 Introduction

As Chapter 1 explained, this thesis aims at contributing to both environmental flow assessments and Integrated Water Resources Management (IWRM). This second chapter explores these two topics to understand what the state-of-the-art is in both fields, what new developments are going on, how human well-being is addressed in the approaches currently used, and what knowledge gaps can be identified. The findings form the starting-point, challenge and constraints for the research on linking human well-being to environmental flow assessments for application in IWRM.

Section 2.2 describes the concept of environmental flows and methods to assess these flows, as well as the link with human well-being. The concept of IWRM and how multiple interests are dealt with is discussed in section 2.3. Section 2.4 discusses the applicability of environmental flow assessment methods in IWRM. Section 2.5 formulates the challenges for this thesis to reduce the identified knowledge gaps. Human well-being itself is discussed in detail in Chapter 3. For the purpose of this chapter, human well-being can be defined as everything important to a person, both physically and mentally.

2.2 Human well-being in environmental flows approaches

2.2.1 Concept & definitions of environmental flows

The term environmental flows or environmental flow requirement is nowadays widely used to indicate the river flow needed to maintain the river ecosystem. The concept emerged in the 1950s, when pollution alone could no longer explain ecosystem degradation (King et al., 1999). Although the degree to which the concept of environmental flows is applied varies considerably among countries, people in over 70 countries worldwide are nowadays aware of the concept, according to a survey carried out by Moore (2004).

The concept evolved along various paths. In the early stages, environmental flows were defined as fixed minimum flows, for example a percentage of mean annual run-off, while nowadays they involve a comprehensive description of various flow characteristics including both high and low flows. Whereas at first environmental flows were determined through black-box methods, with hydrological indices to describe different levels of ecosystem integrity, nowadays links between flows and ecosystem components are increasingly made explicit. The scope of the environmental flows concept has widened as well. In the beginning, environmental flows were determined with the aim of conserving specific species of fish, while nowadays entire ecosystems, including plants and animals living on floodplains and levees, and people depending on the ecosystems for their livelihood and health are included in the concept.

Various definitions of environmental flows can be found in literature. The following definition of environmental flows is adopted in this thesis:

> the proportion of the *original flow regime* of a river required to maintain *specified valued features* of the *river ecosystem (Tharme, 2003)*

The *river flow regime* can be described as the full range of natural intra- and interannual variation of hydrological regimes, and associated characteristics of timing, duration, frequency and rate of change (Richter et al., 1997). This encompasses low flows and floods of various magnitudes and return periods. The underlying assumption is that the condition of the river ecosystem is the result of the prevailing flow regime, and that each of the flow characteristics is relevant for maintaining part of this river ecosystem (Brown & King, 2000). Nevertheless, authors have postulated that certain flow characteristics can be decreased without major effects on the river ecosystem. This allows for abstracting the amounts that are considered superfluous in certain time periods (O'Keeffe, 1999). A common misunderstanding about the environmental flows concept is that it deals with low flows or droughts. Low flows and droughts (See Box 2.1) are both natural hydrological phenomena. Generally, these phenomena are required to maintain the ecosystem in the natural condition. Water scarcity can be defined as a discrepancy between supply of and demand for water. This can be the case in normal flow or even in high flow situations. Changes to the river flow regime may result in structural changes in the frequency, duration and timing of low flows and droughts, and an increase of water scarcity. However, development projects not only lead to a reduction of flow: operation of hydropower reservoirs will, besides reducing high flows, lead to increased low flows. Both reductions and increases of flows, or a change in timing or other parameters, constitute deviations from the natural flow regime and will affect the river ecosystem. The concept of environmental flows aims at a natural flow regime, which does not necessarily mean a larger volume of water then available after an intervention.

> **Box 2.1 Terms used for low availability of water**
>
> **Low flow**
> Seasonal phenomenon, and an integral component of any river. (Smakhtin, 2001)
>
> **Drought**
> Natural event resulting from a less than normal precipitation of an extended period of time (Smakhtin, 2001)
>
> **Water scarcity**
> Discrepancy between supply and demand for water, which can even be the case in normal flow situations

The *river ecosystem* can, for the purpose of this thesis, simply be defined as the part of the ecosystem affected by changes in the river flow regime. More specifically, the river ecosystem, sometimes called riverine ecosystem, has been defined as all components of the landscape that are directly linked to that river and all their life forms, including the source area, the channel from source to sea, riparian area (i.e. the longitudinal riverside strips with vegetation types that are distinct from the general terrestrial landscape), the water in the channel and its physical and chemical nature, associated groundwater in channel and bank areas, wetlands linked either through surface water or groundwater, floodplains, the estuary, and the near-shore marine ecosystem if this is clearly dependent on freshwater inputs (King et al., 1999). It is important to realise that the river ecosystem consists of both the water itself and the flora and fauna depending on this water, but not necessarily in this water.

The *specified valued features* refer to all parts of the ecosystem which should be maintained. This poses river managers for the difficult task of defining the desired condition of a river ecosystem to be maintained. A completely natural flow regime is normally not achievable because of the presence of other water users. The second option is 'minimum degradation' (Brown & King, 2000), making use of the assumptions that some flow characteristics can be reduced without major effects. If this option is not feasible either, the question becomes: how much degradation is acceptable? This issue is discussed in more detail in section 2.4.

Now that the definition of environmental flows for this thesis has been discussed, this section will explain a few other terms which are used to describe more or less the same concept. Moore (2004) lists 40 such terms, with slighter or larger variations in precise meaning. Three commonly used terms are discussed here: minimum flow, instream flow and managed floods.

The term *minimum flow* was one of the first terms to describe the amount of water required in the river to maintain the river ecosystem. For two reasons this term is not much in use anymore: 1) it seems to refer to a fixed minimum instead of to the required natural variation in the flow regime, which can even include zero flows at certain moments, and 2) it seems to refer to the amount of water to maintain a minimal, instead of an optimal ecosystem condition (Stalnaker et al., 1995).

Instream flow is a widely used term, especially in the US, referring to the same concept as environmental flows. A drawback of this term is the apparent focus on the river channel, whereas floodplains and levees should be comprised as well. Brown & King (2003) distinguish environmental flows as a specific type of instream flow, identifying the following instream flows which are not environmental flows: hydropower releases, irrigation releases, flows to enable navigation, flows to dilute pollution, release of wastewater and interbasin transfers.

Managed floods are defined as a controlled release of water from a reservoir to inundate a specific area of floodplain or river delta downstream to restore and maintain ecological processes and natural resources for dependent livelihoods undertaken in collaboration with stakeholders (Acreman et al., 2000; McCartney & Acreman, 2001). This concept can be considered a specific type of environmental flows, focused on the use of the floodplain by local communities, and not so much on biodiversity as such.

To actually assess the environmental flow requirements, various methods have been developed. These environmental flow assessment methods will be discussed in the next section.

2.2.2 Environmental flow assessment methods
Ever since the emergence of the concept of environmental flows, methods are being developed to assess what the environmental flow requirement should be. Together

with the evolution of the concept these methods have evolved. Nowadays, there are over 200 methods, reviewed by various authors (e.g. Karim et al., 1995; Jowett, 1997; Dunbar et al., 1998; King et al., 1999; Tharme, 2003). The methods are referred to as either environmental flow assessment (EFA) methods, or simply environmental flow methods or EF methods.

Each review states that there is not one best EFA method. What method is appropriate depends on a number of issues, such as the availability of data for the studied river, location and extent of the study area, time and financial constraints or the level of confidence required in the final output. Simple methods from the early years of EFA method development may still be useful in specific situations where quick methods are required or where the focus is on specific aquatic species.

Generally, the reviews distinguish four main categories of EFA methods: 'hydrological', 'hydraulic rating', 'habitat simulation' and 'holistic' methods. Other classifications can be found as well (e.g. in Acreman & Dunbar, 2004). *Hydrological methods* make use of statistical analysis of the river flow regime, and assume that a certain deviation from the natural flow regime leads to a certain degradation of the river ecosystem. Examples of commonly used hydrological methods are the Tennant method (Tennant, 1976), and the Range of Variability Approach (RVA) (Richter et al., 1997). This latter method uses 32 hydrological indices, and is the most extensive hydrological method. *Hydraulic rating* and *habitat simulation methods* focus on the habitat requirements of specific species. To represent habitat requirements, hydraulic rating methods use more simple hydraulic parameters than habitat simulation methods. The habitat simulation methods often involve detailed mathematical modelling, and are mainly applied in developed countries in the northern hemisphere (Tharme, 2003). An example of a habitat simulation model is PHABSIM (Stalnaker et al., 1995). *Holistic* methods were first developed in the 1990s aiming at defining flow requirements to protect the entire river ecosystem. Holistic refers to the consideration of the entire ecosystem that depends on the river flow regime, including for example floodplain vegetation. These methods combine various disciplines and can be either based on discussion and expert judgement, or on calculation tools, or on a combination. Holistic methods are the only methods explicitly dealing with social aspects and are discussed in more detail in the next section.

2.2.3 Human well-being in environmental flow assessment methods

Various authors have identified the importance of rivers and associated ecosystems for people, especially for the poor, as was discussed in Chapter 1. Yet, only few environmental flow assessment methods explicitly consider social or livelihood aspects. Three methods which do take such aspects into account are the Building Block Method, DRIFT and the Managed Floods Approach, and are discussed in this section.

Building Block Method

The Building Block Method (BBM) was developed in the 1990s, as one of the first holistic methods, to suit South Africa's need for an environmental flow assessment method which would address the health (structure and functioning) of all components of the riverine ecosystem, instead of focusing, as most of the methods until then did, on selected (aquatic) species (King et al., 2000). The building blocks refer to identified parts of different flow components (low flows, floods with certain magnitude, timing and duration), which are considered to perform a required ecological or geomorphological function.

In addition to the assessment of flow components for the ecosystem itself, a social assessment is carried out to determine the desired future state of the river ecosystem from the viewpoint of local communities. To achieve this, the social assessment of the BBM aims at assessing the importance of a healthy river ecosystem for sustaining rural livelihoods. The focus is on the direct use of natural resources by rural communities. All resources are ranked for their priority and an inventory is made of to what groups of people within the community the products are important. Pollard (2000) recommends that three key issues should receive attention in assessing social implications of flow changes: 1) health related impacts, 2) economic impacts, and 3) changes in the social dynamics of communities through changes in access to resources. Identification of rural communities who may be using the river ecosystem is done through a social survey. After the communities have been identified, participatory rural appraisal methods (PRA) are used for carrying out the steps of the assessment together with the community members. Box 2.2 shows an outline of the different steps of the social assessment of the BBM.

Box 2.2 Overall framework of the social assessment of the Building Block Method (Source: Pollard, 2002)

---

1. General Resource inventory
    Review project objectives and research approach with participants
    - Identify which riverine resources are used
    - Identify who uses them

2. Focus group discussions to discuss natural resources used
    a. Define resource attributes
    - Prioritise the relative importance of each resource or use
    - Describe the location and extent of resource (if appropriate)
    b. Establishment of the link between the resource and flow
    - Describe seasonality-of-use (and hence discharges) that are important for resource maintenance
    - Describe critical water levels associated with each resource
    - Examine changes in resource availability/use over time.

3. Define a Desired Future State (DFS) of the river
    Collate the above information into a shared vision for a DFS

4. Integrate findings into Environmental Flow Assessment
    Cross-check results with biophysical specialists and integrate into EFA.

The main weakness of the BBM is the fact that the method is prescriptive (King et al., 2003): a desired river condition is specified upfront, after which the flow regime required to achieve this is described. The outputs are directed at justifying a single flow regime, while the effects of not meeting the required flow regime are not assessed. The fact that the social assessment leads to a desired future state, but does not provide information on the social effects of not meeting the flows required to maintain the identified state of the river ecosystem, is also already mentioned in the manual of the BBM itself (King et al., 2000).

DRIFT

The method DRIFT (acronym for Downstream Response to Instream Flow Transformations) is essentially a system for managing data and knowledge. The main

Chapter 2

rationale is that different parts of the flow regime elicit different responses from the river ecosystem (Brown & King, 2000). The method consists of four modules (Figure 2.1):

1. a biophysical module;
2. a sociological module;
3. a scenario development module; and
4. an economic module.

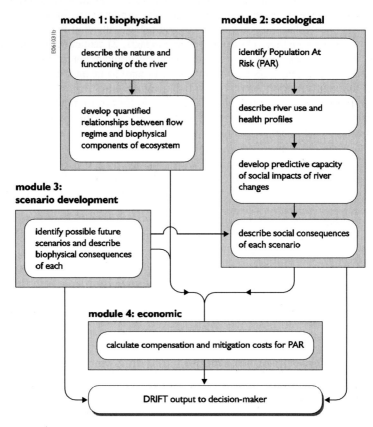

Figure 2.1 The four modules of the DRIFT method (source: King, Brown & Sabet, 2003)

In the *biophysical module*, isolated responses of ecosystem components to changes in the various flow components are estimated. This requires at first an analysis of the hydrological regime and the identification and isolation of various flow components such as wet-season and dry-season low flows, and small and large floods from the long-term hydrological record. For every change in flow regime, changes in ecosystem components are estimated with a plus or minus sign to indicate positive or negative change and with a number of zero to five to indicate the severity of the change. To deal with uncertainty, ranges of change can be used. All estimated responses are combined in a database.

In the third module, *scenario development*, possible future flow regimes are defined, expressed in terms of the identified flow regime components. The database can then be used to find the estimated ecosystem change resulting from each of the scenarios. Socio-economic results are subsequently estimated for the resulting ecosystem condition.

The *sociological* module does not consist of a database with responses to isolated flow regime changes. Instead, the social consequences are estimated based on the biophysical responses to the scenarios. This module has the same components as the social assessment of the Building Block Method: identification of population at risk (PAR), identification of natural resources used and health threats.

The main advantage of DRIFT compared to the BBM is that the social assessment does not stop with identifying the population and the ecosystem components which they consider important. After the biophysical responses to the identified scenarios have been established, the results are used to determine two types of social consequences (King et al., 2003):

- changes in resources;
- changes in health profile and overall quality of life.

Changes in resources are determined in a quantitative way, and form the input for the *economic module*. In this module economic losses are assessed in monetary terms.

Changes in health profile and overall quality of life are described in a qualitative way. For this latter assessment no specific method is provided, but a number of considerations on this type of impact are given (King et al., 2003):

- different river uses are linked to different illnesses;
- different river changes will affect health in different ways;
- frequency of river use heightens the chance of water-related illnesses; and
- even with a high frequency of river use, a known health threat can be mild and thus of low impact.

- The quantified ecosystem responses and economic impacts and the qualitative description of social impacts constitute the environmental flows information that goes to the decision-maker. The quantification of the impacts on subsistence users is identified as one of the issues on which further development of the method should focus (King et al., 2003).

Managed Floods Approach

The Managed Floods Approach, developed by Acreman *et al.* (2000) focuses on the livelihoods of communities using the floodplains. Floodplain inundation is here considered the key hydrological characteristic to maintain the floodplain ecosystem features that local communities depend upon. The approach was developed as a mitigation measure at dams (McCartney & Acreman, 2001). Participatory rural appraisal methods are advocated to involve stakeholders, and to assess the human aspirations. The approach consists of 10 steps divided over three levels (Figure 2.2): 1) feasibility, 2) design and 3) implementation. For this thesis, level 2, design, is the most relevant. This level consists of steps 3 through 7:

Chapter 2

3. develop full stakeholder participation and technical expertise;
4. define links between floods and the ecosystem;
5. define flood release options;
6. assess impacts of flood options;
7. select the best flood option.

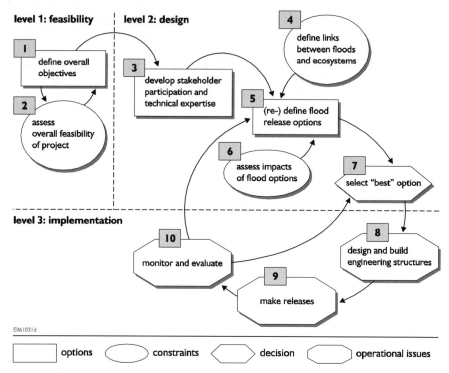

Figure 2.2 Steps of the Managed Floods Approach (Source Acreman *et al.*, 2000)

In step 3 and 6, not only the downstream stakeholders and effects are included, but also the effects on the beneficiaries of the reservoir: people benefiting from irrigated agriculture, domestic water supply, hydropower or industries. The stakeholders' objectives are described in terms of the livelihood outcomes of the Sustainable Livelihoods Framework of the DFID (Scoones, 1998, see Chapter 3 of this thesis), which are:

- increased income;
- enhanced environmental sustainability;
- improved food security;
- reduced (or not enhanced) vulnerability; and
- increased well-being – which must be defined in a participatory way and may include intangibles like access to social networks and maintenance of culturally important seasonal cycles.

The Managed Floods Approach provides a direct link between different flood characteristics, the resources people use and the outcomes for their livelihoods, and can in this way assess impacts of various flood release options. The method itself is not an environmental flow assessment method, but will make use of appropriate environmental flow assessment methods to assess the link between the flow regime and the ecosystem condition. The available literature does not specify exactly how the relationships between floods and livelihoods will be assessed, but focuses on the different steps to take in a complete program of assessing flood requirements, implementation and monitoring and redefining flood release options.

2.2.4 Conclusions on human well-being in environmental flows approaches

The environmental flows concept has over the years evolved from a fixed minimum discharge, towards a variation in low and high flows both within and over the years, and from a focus on instream species, towards the consideration of all river flow dependent flora and fauna. Various environmental flow assessment methods have been developed subsequently to quantify the desired flow regime and related ecosystem condition.

Holistic methods include, or have the potential to include, human well-being. Examples are the Building Block Method and DRIFT. These methods include steps to identify potentially affected communities and assess the use of resources and ecosystem goods and services. In the BBM and in DRIFT, human well-being is linked to ecosystem condition. The Managed Floods Approach may incorporate holistic EFA methods in linking floods and human well-being, but also directly relates livelihoods to flood characteristics such as timing, frequency and magnitude of floodplain inundation.

Despite the fact that human well-being is conceptually included in the BBM, DRIFT and the Managed Floods Approach, literature on the Managed Floods Approach does not specify how relationships between floods and livelihoods can be quantified and the social module of the BBM and DRIFT remain qualitative descriptions. As the developers of DRIFT indicate themselves, the quantification of the social impacts requires further elaboration. It can be concluded that there is a need for a transparent approach to link river flows and ecosystem condition to human well-being in a way in which this relationship can be quantified.

2.3 Human well-being in Integrated Water Resources Management

2.3.1 Concept & definitions of Integrated Water Resources Management

The concept of Integrated Water Resources Management was introduced around the 1980s to change the, at that time prevailing, sector-wise approach in water management. To protect and manage water bodies, people realised that it was necessary to consider quantity, quality and ecological issues, as well as the interests of the people who make use of or are threatened by the water. This requires involvement of all institutions which represent certain groups of people or are responsible for specific components of the water resources system. This approach is often visualised by the interrelationships of the three sub-systems which constitute the water resources system (Figure 2.3)(Loucks & Van Beek, 2006):

Chapter 2

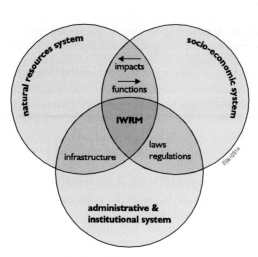

Figure 2.3 Schematic presentation of the interrelationships between the natural resources, socio-economic and administrative & institutional systems

- *natural resources subsystem*: the system of rivers, lakes, groundwater aquifers including its functions for the ecosystem, and the infrastructure required to use the water resources.
- *socio-economic subsystem*: water use and other water related human activities.
- *administrative & institutional subsystem*: system of administration, legislation and regulation, where the decision, planning and operational management processes take place.

Following the evolution of IWRM concepts, actual water resources management has over the years developed from a technical approach, with a technical analysis and technical measures, towards a process which aims at including all stakeholders and reaching consensus on the desired status of the water resources system and the measures to achieve this. The definition of IWRM by the Global Water Partnership (GWP) (2000) reflects this:

> *IWRM is a process which promotes the co-ordinated development and management of water, land and related resources, in order to maximize the resultant economic and social welfare in an equitable manner without compromising the sustainability of vital ecosystems.*

At the basis of this definition of IWRM lie the so-called Dublin principles (Box 2.3). These four principles, formulated at the International Conference on Water and the Environment in Dublin, 1992, are nowadays commonly accepted amongst the international community as the guiding principles underpinning IWRM (GWP, 2000). To achieve sustainable management of river basins, development is required both in the institutional area and in providing analytical support for planning and operational management (Mostert et al., 2000). In this thesis, the focus is on contributing to the analytical support. The following sections discuss how decision-making in water resources management takes place and what information, and analytical support to provide this information, is required.

Box 2.3 Dublin principles (GWP, 2000)

> Dublin Principles (GWP, 2000)
> - Fresh water is a finite and vulnerable resource, essential to sustain life, development and the environment.
> - Water development and management should be based on a participatory approach, involving users, planners and policy-makers at all levels.
> - Women play a central part in the provision, management and safeguarding of water.
> - Water has an economic value in all its competing uses and should be recognised as an economic good.

2.3.2 Systems analysis and decision-making in IWRM

The growing number of users, the desired rise in living-standards, and the need to reduce environmental degradation have led to high pressures on the water resources system. Increased use of water resources by some will reduce the opportunities for use by others. To consider the interests of all stakeholders, as promoted in IWRM, it is necessary to analyse the complex water resource system in its entirety. Systems analysis provides a structured framework for analysing complex problems, and is often applied in IWRM. According to Miser and Quade (1985) problems can be reduced to evaluating the efficiency of alternative means for a designated set of objectives. This means that the following questions are relevant:

- What are the objectives?
- What are the alternatives for achieving these objectives?
- How should the alternatives be ranked?

This section describes the analytical process of systems analysis in IWRM. This process is discussed in detail in Loucks & Van Beek (2006).

Before the water resources system can be analysed, the objective for managing the system needs to be established. Subsequently, it will be assessed whether or not these objectives are met at present and at the time horizon defined for the management of the water resources system. When the objectives are not sufficiently met, potential strategies to improve the situation can be developed and analysed. *Strategies* are here understood as single measures or combinations of measures. Water resources management measures are often divided into three groups: supply-oriented, demand-oriented or management measures. Constructing a new reservoir is an example of a supply oriented measure. A demand-oriented measure could be the shift to crops which require less water. Management measures could be a change in operation rules for an existing reservoir or the establishment of new laws to regulate ecosystem use. To deal with uncertainty about future developments the analysis can be done under various scenarios. In this thesis, *scenarios* are changes in the system, which are outside the range of influence of the decision-makers involved, and this way are distinguished from strategies, which are changes a decision-maker can influence. Examples of scenarios are climate change, developments in another administrative region or country, population growth or variations in the global trade market. A water resources systems analysis is preferably carried out at the basin level.

To measure the extent to which the water resource management strategies contribute to achieving the water resources management objectives, it is necessary to define operational criteria. Operational refers to the fact that the criteria can be given a unambiguous value, either in a quantitative or qualitative way. Criteria are related to indicators. UN (2003) defines an indicator as a single data, variable or value that can be used to describe the state of a system or process. A criterion defines in what range, or above or below what threshold, the value of the indicator should be in order to evaluate whether objectives are met. The effects of alternative strategies can then be expressed in terms of these indicators and criteria. Typically, the criteria scores for all analysed strategies and scenarios will be displayed on score-cards, together with information on cost and feasibility of the strategies. Such score-cards provide the basic information based on which decisions can be taken.

In the process of IWRM, the task is to find a solution which is acceptable to all stakeholders. In many cases, it will not be possible to derive one optimal solution which will fulfil the requirements of all water users. Trade-offs need to be made, preferably in a participatory process of negotiation. For systems analysis this means that not only the total performance of the system needs to be assessed, but that information needs to be generated on who benefits or looses as a result of a strategy. The identified criteria need to represent the interests of the stakeholder groups.

The selection of the preferred strategy will be done by the decision-makers. This group of decision-makers should preferably include representatives of all relevant stakeholder groups. The decision-makers should be involved from the beginning of the analysis: they should formulate the objectives and criteria for management of the water resources system, and should indicate what measures could be feasible. Systems analysis as described here focuses on the analysis of various possible strategies in order to aid decision-makers. The process of actually making the decision, implementing the selected strategy, evaluating the real consequences, and designing new strategies is not part of the systems analysis.

Assessing the problems for water resources management in the present and the future situation requires a detailed understanding of the water resources system. This understanding is also required to estimate the impacts of water resources management strategies under various scenarios. Because of the complexity of the water resources system, obtaining such an understanding of the system often involves computer modelling. The most used tools for such a systems analysis are generic water-balance and -allocation models which can be applied at various spatial scales including the basin scale. Examples of such models are RIBASIM (WL | Delft Hydraulics, 2004), Mike-basin (DHI, 2003) and WEAP (SEI, 2001). These models contain modules to include water demands for agriculture, hydropower, drinking water etc., together with various types of infrastructure. In a model application, water resources management strategies can be simulated at time scales of days to months.
Based on the above description, the systems analysis approach can be summarised as consisting of the following steps:

- identifying and consulting stakeholders;
- formulating objectives and indicators;
- developing scenarios;
- data collection and modelling of the water resources systems;
- assessing the current and the future state of the system: are objectives met, or can problems be identified?
- designing measures and strategies to reduce the identified problems;
- evaluating the possible impacts of the strategies;
- formulating the impacts of the strategies in terms of the criteria.

The sequence of these steps is shown in a schematic way in Figure 2.4. Although systems analysis is described here as a sequential process, in reality various iterations will be required, especially to adjust measures and strategies based on the results.

2.3.3 Human well-being related to ecosystems in IWRM

All developments are eventually aimed at improving human well-being. According to its definition, the goal of IWRM is the maximisation of social welfare, which can be considered synonymous to human well-being. However, this implicit understanding is often not translated into explicit criteria to measure the performance of the water resources system. As was discussed in Chapter 1, literature on IWRM lists three main criteria that IWRM should comply with (ESCAP, 2000; GWP, 2000):

- *Economic efficiency* in water use: because of the increasing scarcity of water and financial resources, the finite and vulnerable nature of water as a resource, and the increasing demands upon it, water must be used with maximum possible efficiency.
- *Equity*: the basic right for all people to have access to water of adequate quantity and quality for the sustenance of human well-being must be universally recognized.
- *Environmental and ecological sustainability*: the present use of the resource should be managed in a way that does not undermine the life-support system thereby compromising use by future generations of the same resource.

Human well-being is explicitly mentioned in the second criterion, with respect to access to water for sustenance. However, in the sectoral approach in water resources allocation, the contribution of the various sectors towards these criteria is not always given due consideration. In the World Water Development report (United Nations, 2003) five water demand sectors are identified[1]: 1) basic human needs, 2) food, 3) nature, 4) industries, and 5) energy. In this division, basic human needs refer to water for drinking and sanitation, food is purely related to agriculture, nature is important for sustainability, to protect the resource base for future generations, and industry and energy are important for regional and global economies. This division in sectors is misleading. Allocation of water to irrigated agriculture is well-known to

---

[1] In the recently published second World Water Development Report ((United Nations, 2006), wetlands are included in a chapter on Water for Food, Agriculture and Rural Livelihoods. However, for the message this section wants to convey, it was believed justified to maintain the division of the first World Water Development Report.

Chapter 2

provide much more than just food. Various authors have stressed the multiple-use of irrigation water: for growing home-garden crops which promotes gender-equity, for water for additional income-generating activities such as brick-making, and simply for domestic purposes improving human health (Meinzen-Dick & Bakker, 1999; Barker et al., 2000). In a similar way, nature is not only relevant to protect the resource base for future generations. Natural ecosystems are of utmost importance to the livelihoods of the poorest people in this world today (Silvius et al., 2000; Mainka et al., 2005).

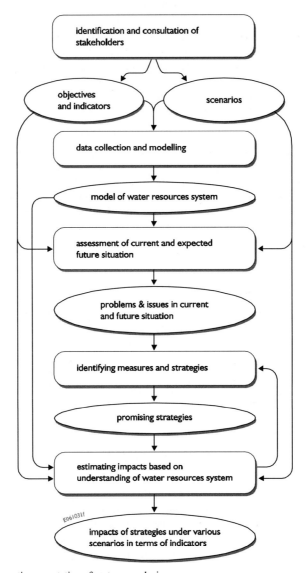

Figure 2.4 Schematic presentation of systems analysis

As the two largest consumers of water, the debate is often about the balancing of the water needs of irrigated agriculture and natural ecosystems (Rijsberman & Molden,

2001; Molden & De Fraiture, 2004). In this debate as well, irrigated agriculture is seen as providing food and livelihoods to rural communities, whereas the value of nature lies mainly in the protection of future resources. This is reflected in GWP's (2000) formulation of the challenge of IWRM as 'striking a balance between the use of the resources as a basis for the livelihood of the world's increasing population and the protection and conservation of the resource to sustain its functions and characteristics'. Although this challenge indeed exists, it gives a rather black and white picture, where people either withdraw water for their own benefit or leave it in the river to sustain nature for future generations.

Various authors have stressed the need to move away from the sectoral approach and to put human livelihoods at the centre of water resources decision-making (Rijsberman & Molden, 2001; Merrey et al., 2005). These authors express the need for a comprehensive set of criteria. With respect to the three IWRM-criteria, this will mean a further specification of social equity. Also, it requires that the contribution of allocating water over different water use sectors to these criteria is made explicit. Therefore, a modified division in water use sectors is suggested in Figure 2.5. The sector 'food' is replaced by agriculture, which provides food and other benefits, while food has become part of the equity criterion to which both agriculture and nature contribute. Moreover, because of the various goods and services nature provides, allocation of water to nature can also be an economic efficient solution to water resources management problems. Basic human needs are considered a value instead of a sector and therefore part of the social equity criterion, while the sector that water is allocated to is replaced by 'domestic water supply'. The link from the allocation of water through natural ecosystems towards the expanded social equity criterion is central in this thesis.

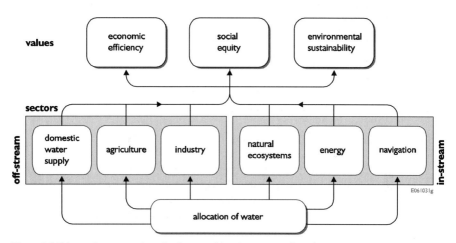

Figure 2.5 Schematic presentation of values resulting from water allocation

In the water balance and –allocation models used to analyse water resources systems, environmental flows can generally be included through specifying minimum flows for certain time periods and river stretches. However, this option does not accommodate requirements in terms of frequencies of flood events. Moreover, without quantified relationships between these flood events and human well-being the changes in social equity resulting from alternative water resources management strategies cannot be assessed. Further development of these models will benefit from insight in what is involved in quantifying ecosystem condition and associated well-being values.

2.3.4 Conclusions on human well-being in Integrated Water Resources Management

Human well-being and social equity are conceptually included in Integrated Water Resources Management. Yet, in practice, the assessment of equity is focused too narrowly on drinking water and sanitation and livelihoods related to agriculture, and neglects the contribution of the natural ecosystem. There is a need to consider the contribution of the different water use sectors to the IWRM criteria in a wider perspective and to properly recognise the relationship between water allocation for nature and the criteria of economic efficiency and social equity. When this relationship is understood, water balance models can be extended to include river ecosystems as one of the water demand sectors. This requires appropriate indicators and quantifiable links. Quantification is important in order to ensure that the value of nature can be included in score-cards, next to the benefits from irrigated agriculture, domestic water supply, and the other sectors. Such information will enhance informed decision-making in water resources management.

2.4 Environmental flows in IWRM

2.4.1 Impediments to environmental flows application and implementation

Environmental flow assessment methods were developed with the specific aim to improve the management of rivers. Yet, application of the concept and actual implementation of environmental flows is not yet common. This was one of the main conclusions of the 2002 Cape Town conference on environmental flows. Brown and King (2003) identify a number of reasons which impede the application and implementation of environmental flows:

Problems with application:

- lack of data and understanding of hydrology-ecology linkages;
- lack of specialists in developing countries.

Problems with implementation:

- unwillingness or inability to incorporate innovative, and possibly more expensive, release mechanisms into dams for environmental releases;
- perception that too much water is being requested;
- lack of political or legislative pressure to implement the environmental flows (usually because other demands were seen as more important);
- 'last minute' or post-hoc flow assessments that are commissioned after most (if not all) of the major decisions about the design and cost of the development, and the allocation of water, have already been made;
- reluctance to move away from established practices.

The problems with application are related to capability and knowledge (including data-access) of researchers and consultants, while the other problems are related to the mind-set of decision-makers and river basin managers. If these decision-makers and river basin managers can be convinced of the need for environmental flows, legislation and implementation will follow.

A third type of reason, which may explain problems with assessing environmental flows, is the gap between approaches in the fields of environmental flows and Integrated Water Resources Management. Traditionally, water resources development and management is the domain of engineers, while environmental flow assessments generally involve ecologists and social scientists with various specialisations. Even with a political will and capable specialists to assess environmental flows, good communication between the specialists of these two fields is required in order to facilitate a linkage of approaches and to make results comparable. Only then truly informed decision-making can take place.

2.4.2   Suitability of environmental flow methods for IWRM systems analysis

A first step to decrease the gap between environmental flows and Integrated Water Resources Management is to evaluate which of the environmental flow assessment methods match best with the systems analysis approach outlined in the previous section. A useful division for this discussion is the division into 'objective-based' versus 'scenario-based' methods (Dunbar et al., 2004). Other authors use other terms for the same division: 'prescriptive' versus 'interactive' (Brown & King, 2003), or 'standard-setting' versus 'incremental' (Stalnaker et al., 1995).

The first holistic EFA methods were *'objective-based'*, which means that the desired state of the river ecosystem is defined upfront, and that subsequently the amount of river water required to maintain or restore the ecosystem condition is assessed. Such an approach requires a legal framework to enforce the implementation of the assessed environmental flows. In South Africa, a National Water Act was established in 1998, to provide such a legal framework. This Water Act states that allocation of water for commercial use can only be done after a certain amount of water has been allocated to basic human needs and to the environment, the so-called Reserve. A system of 'Ecological Management Classes' was developed to assign current and future classes to rivers (Box 2.4), in order to facilitate setting objectives for the ecosystem. The social assessment of the BBM was developed to help define the desired future state for a particular river.

When a legal framework is not in place, actual implementation of predefined environmental flows for ambitious ecosystem conditions is highly unlikely. Requesting a fixed flow regime carries the risk that it is considered too much and that environmental flows are not considered at all (Stalnaker et al., 1995; Brown & King, 2003). Moreover, it is not up to the researcher, nor to the subsistence communities alone, to decide what the desired future state of an ecosystem is. As a solution to these problems the *'scenario-based'* approaches were developed.

Box 2.4 South African National Water Act (Source: (O'Keeffe & Louw, 2000))

South African National Water Act
The new South African National Water Act came in effect in 1998. An important component of this Water Act is the safeguarding of the Reserve. The Reserve consists of two parts: 1. water for Basic Needs, and 2. the Environmental Reserve. These are the only water rights, and only the water exceeding this amount is available for commercial use. Ecosystems can be maintained at different levels of condition. To facilitate the process of setting the reserve, Ecological Management Classes were defined to describe the present and desired river conditions.

Ecological Management Classes:
Class A -   Close to natural conditions
Class B -   Largely natural with few modifications
Class C -   Moderately modified
Class D -   Largely modified
Class E -   Seriously modified; no longer providing sustainable services
Class F -   Critically modified; no longer providing sustainable services

The six classes can be used for describing the current state, and class A through D can be used to define the desired future state. The Ecological Management Class is set for each section of river in a procedure which takes account of the technical assessments of the specialist ecologists, and the wishes of the stakeholders, but is eventually the responsibility of the Minister of Water Affairs.

Between a natural ecosystem which requires the complete natural flow regime, and a largely degraded ecosystem resulting from highly reduced flows, a continuum of possible flow regimes and associated ecosystem conditions exists. King, Brown & Sabet (2003) argue that what the desired ecosystem condition to be maintained is, needs to be determined through negotiation. Negotiation requires insight in the effects of alternative flow regimes, which implies that a single flow requirement does not suffice. DRIFT and the Managed Floods Approach are both examples of scenario-based approaches. What the desired ecosystem condition and related flow regime is is not defined upfront but is the result of the decision-making and negotiation process, in which also the effects of (not) changing flow regimes for upstream users are considered.

Since the approach in IWRM focuses on the balancing of interests and the analysis of the effects of different options, the scenario-based environmental flow assessment methods seem most appropriate for use in an IWRM study. Environmental flows fit perfectly in the approach depicted for analysis of water allocation strategies as the water demand for ecosystems. An assessment of the contribution of alternative environmental flows to the objectives for the water resources system will clarify the importance of the river ecosystem. The ecosystem condition and associated flow regime that will be decided upon through a negotiated trade-off constitutes the actual environmental flow requirement. This flow requirement can then become a fixed requirement in models for the operational management of the (sub-) basin.

2.4.3 Conclusions on environmental flows in IWRM
Although assessments of environmental flows are increasingly widespread, implementation of the assessed flows is not yet common. Lack of knowledge on how results of environmental flow assessments can be included in IWRM systems analysis can be identified as one of various possible reasons for this. To include environmental flows in IWRM, the scenario-based approaches are the most suitable, but further

development is required to facilitate the assessment of to what extent the environmental flows contribute to achieving water resources management objectives.

2.5 Conclusions: Challenge for this thesis

To understand what developments with respect to the assessment of human well-being values of environmental flows are required, this chapter discussed the approaches currently used in environmental flow assessments and IWRM, and with this answered research question 1.

Of the various types of environmental flow assessment methods, the holistic methods are the only methods which include human well-being in their concept. Holistic methods are also the only methods which explicitly assess the relationship between river flows and various resources, which could possibly be of importance to people.

To link environmental flows to IWRM, it is important that the environmental flow methods and the results fit into the systems analysis approach in IWRM. The 'scenario-based' environmental flow assessment methods are considered highly suitable to make the link with systems analysis. An example of such a method is the environmental flow assessment method DRIFT.

However, simply linking holistic, scenario-based environmental flow methods with the IWRM systems analysis approach, does not guarantee that human well-being related to river ecosystems is given due attention in IWRM decision-making. Two gaps can be identified: 1) the social part of the environmental flow assessment methods does not provide quantifiable transparent relationships, and 2) in IWRM the social equity criterion does not include all aspects of human well-being, but tends to focus on water for drinking and sanitation. In the division of water demand sectors, only the domestic water supply and agricultural sector are given proper recognition for their contribution to human well-being, while the contribution of natural ecosystems is not given much attention. Therefore, this chapter has suggested a new division in water use sectors and has redefined the social equity criterion, in order to clarify that various water use sectors, including natural ecosystems, can contribute to enhancing social equity.

Although the concept of environmental flows is widespread, actual implementation in practical situations lags behind. Various reasons for this are suggested of both technical and political character. Linking environmental flow assessments to systems analysis and including values for human well-being will contribute to a more comprehensive assessment of impacts of water resources management strategies, and enlarge the chances of environmental flows being implemented, or at least being given fair consideration.

The challenge for this thesis is therefore to develop 1) suitable indicators for human well-being and social equity that can be used as decision-making criteria, and 2) to derive an approach to transparently quantify the contribution of river ecosystems to human well-being in terms of these indicators. The developed indicators and approach should fit in the approaches of both holistic, scenario-based environmental flow methods and water resources systems analysis. This is the topic of Chapter 3.

3 A conceptual model for the relationship between human well-being and environmental flows

3.1 Introduction

From the analysis of current approaches in environmental flows and Integrated Water Resources Management it was found that an approach to quantify changes in human well-being and social equity in terms of appropriate indicators was missing. To develop such an approach a clear conceptual understanding of human well-being and the relationships with the river ecosystem and river flow regime is required. This chapter discusses these three components: human well-being, the river ecosystem and the river flow regime, and constructs a conceptual model of their interrelationships. Although the focus of this thesis is on the relationship between river ecosystems and human well-being, an understanding of the relationships between the river ecosystem and the river flow regime is required to enable the quantification of impacts of changed flow regimes on human well-being.

The development of the conceptual model is discussed in sections 3.2 till 3.6. Section 3.7 discusses how the conceptual model can be applied in IWRM analyses to quantify changes in social equity as a result of water resources management strategies, resulting in a stepwise approach. Application of the model in two case studies in subsequent chapters tests the suitability of the developed model in practical applications.

3.2 Development of a conceptual model linking human well-being, the river ecosystem and the river flow regime

Conceptual models, typically consisting of blocks and arrows, are useful to provide a complete understanding of various relationships between components of a system. The starting point for the development of the conceptual model for this thesis will be a simple block-arrow model consisting of three blocks (Figure 3.1):

- human well-being;
- the river ecosystem; and
- the river flow regime.

Arrows can connect the blocks in two directions: 1) from bottom to top indicating causal relationships: the river flow regime affects the river ecosystem, and the river ecosystem affects human well-being, or 2) from top to bottom indicating standard-setting relationships: a certain level of human well-being poses requirements to the ecosystem which poses requirements to the river flow regime.

Chapter 3

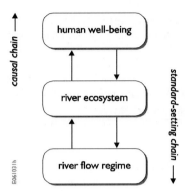

Figure 3.1 Relationship between river flow regime, river ecosystem and human well-being

In the evaluation of strategies in IWRM planning studies, the direction chosen is normally the direction of the causal relationships: strategies are formulated, the system is simulated and impacts on various system components are calculated. For this reason, the 'scenario-based' EFA methods were considered more suitable in IWRM than the 'objective-based' approaches. The main problem with standard-setting chains in planning studies is the need to define upfront what the desired state of the system is, while in practice the results of the analysis will support decision-makers in obtaining an understanding of what they desire the system to be. Although the causal chain is selected as the most suitable one for this thesis, in theory both directions always apply.

The following sections each elaborate one of the three components of the basic model and the link with the other components. In each section this leads to a further extension of the model.

3.3 Human well-being

3.3.1 Components of human well-being

Well-being is regarded in this thesis as encompassing term for everything important to peoples' life. Well-being is a broad concept and its exact meaning is difficult to capture, or measure. A lot of literature on well-being is written in relation to poverty alleviation and socio-economic development. Various authors have stressed the importance of letting the poor define what well-being means to them and what makes that they are poor (Chambers, 1995; Narayan, 1999). Assessments carried out according to this concept revealed that poverty has many dimensions, related for example to access to resources, power, health, or independence.

To quantify changes in well-being, it is necessary to unravel well-being into the various components that constitute well-being. Then, indicators can be defined for each of these components to describe a state of well-being. Literature reveals that unravelling well-being and identifying indicators is difficult, because it is often hard to distinguish between well-being itself and contributions or requirements for well-being, such as income. This issue is referred to as the issue of distinguishing between 'means' and 'ends'(Lok-Dessalien, 1999). While income is often considered an end, it can also be considered a means to obtain nutritious foods and good health. Health is not only an end, but also a means to participate in labour and social activities.

The concept of 'basic needs' defines poverty as deprivation of requirements. This includes also non-monetary income, publicly provided services, potable water and sanitation facilities, employment opportunities and even opportunities for community participation. Because also in this approach it is sometimes difficult to distinguish means and ends, basic needs are sometimes referred to as 'indirect ends' (Lok-Dessalien, 1999).

The Sustainable Livelihoods (Ashley & Carney, 1999) approach combines assets (means) and outcomes (ends) in a single approach. The concept of sustainable livelihoods was introduced in the 1980s to approach poverty alleviation and rural development in a holistic way. Chambers and Conway (1991) define a livelihood as "a means of living, and the capabilities, assets (stores, resources, claims and access) and activities required for it. A livelihood is considered to be sustainable when it can cope with and recover from stress and shocks, maintain or enhance its capabilities and assets, and provide sustainable livelihood opportunities for the next generation; and which contributes net benefits to other livelihoods at the local and global levels and in the short and long term". In the Sustainable Livelihoods Framework (SLF) that was developed from this concept, various types of capital (natural capital, financial capital, physical capital, human capital and social capital) are included as means that can be used through various livelihoods strategies to improve the outcomes for a household's livelihood (Scoones, 1998). Ashley and Carney (1999) list the following livelihood outcomes to achieve: more income, increased well-being, reduced vulnerability, improved food security and a more sustainable use of the natural resource base. In the Sustainable Livelihoods Framework, well-being is mentioned as one of the livelihood outcomes, focused on a mental state. Both the assets and the outcomes are sometimes used as indicators for impacts of changes in water resources management (see e.g. Acreman et al., 2000).

Instead of focusing on well-being itself, Sen (1993) proposed to focus on 'capabilities', which he defines as "a person's ability to do valuable acts or reach valuable states of being". According to UNDP, central to improving human well-being are the choices people have. Development is aimed at enlarging these choices by building human capabilities. As basic capabilities they identify the capability "to lead a long and healthy life, to be knowledgeable, to have access to resources needed for a decent standard of living and to be able to participate in the life of the community"(UNDP, 2005). Capabilities are generally referred to as ends.

Well-being is also assessed through well-known composite indices, such as the Human Development Index or the Human Poverty Index. The first three of the capabilities: longevity, literacy and income are combined in the Human Development Index (HDI), and all four of them, including participation in community life, in the Human Poverty Index (HPI). The difference between the two indices lies in the fact that the HDI calculates an average value for a country, while the HPI focuses directly on the number of people living in deprivation. Such composite indicators may be useful to measure general development over time and to compare development levels of countries, but are less suitable at smaller scales or in water resources planning studies which require information on links between measures and effects (Molle &

Mollinga, 2003). The composite indicators are based on measured changes, and because there are no explicit links between changes in peoples' life and the changes in the components of the indices, the HDI and HPI cannot be used to estimate changes in human well-being.

Summarising, well-being can be characterised as consisting of many components. Identifying the components important to a community is the first step in identifying indicators for human well-being. Composite indices are less suitable for the purpose of this thesis because they don't show what components of well-being are changed for various groups of people. The basic needs concept is useful to link well-being to water resources management. Because of inter-linkages between various components of well-being it is sometimes difficult to define what means are and what ends. The following sections will discuss the complex and dynamic character of well-being in more detail, after which well-being will defined for the purpose of this thesis.

3.3.2 Complex character of human well-being
From the discussion of the components of well-being, it has become clear that not only various components can be identified, but also that many of these components are linked. In their overview of links between water and poverty in various cases, Hussain and Giordano (2004) assert that peoples' health improves as a result of an increase in income: more income generally leads to a higher food intake and to the ability to take measures to prevent disease as well as to treat diseases when they occur. Changes in income are also known to have led to social disruption and increased violence with impacts on physical health and mental well-being (Narayan, 1999).

3.3.3 Dynamic character of human well-being
The socio-economic system is not only complex, but also dynamic. This dynamic nature is also mentioned as important characteristic of the Sustainable Livelihoods Approach (Chambers, 1995; Ashley & Carney, 1999). Especially the poor have no choice but to adjust their livelihood strategies to changed circumstances. This may either improve or worsen their well-being. A strategy of alternative income generation may result in an increase in income, but possibly with negative side effects, such as increased insecurity or health risks. With respect to the assessment of well-being impacts, this means that one needs to be careful while developing and applying ecosystem-poverty relationships. When, for example, fish stocks decline, the income of a fisherman will in first instance decline as well. Depending on the vulnerability of the household, the household may enter a vicious circle in which their poverty increases. However, it is also possible that the fisherman or other household members will find alternative employment or that the household is looked after by others.

3.3.4 Defining human well-being for this thesis
Well-being can now be understood as a dynamic concept, consisting of various linked components. For this thesis, this means that it is important to recognize that changes in the river ecosystem have not only a direct impact on for example income, but may lead subsequently to reduced health, and for example a reduced self-esteem and related mental well-being state. The socio-economic sub-system of the water resources system is highly complex and dynamic. To structure this complex system, this thesis proposes to distinguish well-being components which are directly related to

water or ecosystems as first order well-being components, and refer to subsequent changes in mainly mental well-being as changes in second order components.

Mouafo *et al.* (2002) describe a good example of first and second order impacts resulting from ecosystem degradation downstream of the Maga dam on the Logone river in Cameroon. Impoverishment of the local population was a direct impact of the dam through the collapse of fisheries, the decline of pasture-dependent livestock and the reduction of recession agriculture. These were the first order impacts. Subsequently, various second order impacts were observed: the decreased living circumstances led to a growing insecurity resulting from robbery and disputes between various interest groups. This induced migration of local communities which led to a dislocation of community life.

First order components

To describe *first order* components or values two physiological components and one psychological component of well-being are identified:

- Physiological well-being:
    - income & food;
    - health.
- Psychological well-being:
    - perception & experience related to the environment.

Income & food

With respect to the river ecosystem, income and food represent virtually all products which people obtain from the river ecosystem, except water for domestic use. In rural communities income and food are often hard to separate: generally people take from their yield what they need for themselves and sell excess products. Income can be considered a substitute for various products: with money goods can be purchased from other peoples' yield. For this reason, under income and food we include also free materials taken from the river ecosystem for other purposes than income and food, for example for house construction, household utilities or fuel wood. Reduction of the availability of these resources would mean that part of the income is required to purchase the products from another source.

Health

The WHO (1999) describes health as a complete state of mental, physical and social well-being, not merely the absence of disease or infirmity. In this thesis, however, health refers to physical well-being, since mental and social well-being are included in other components. Besides being related to income and food, peoples' health can also be directly influenced by changes in the ecosystem and the flow regime. People may obtain water for domestic use, mainly for drinking and sanitary purposes, from the river ecosystem. The ecosystem may play a role in purifying the water leading to a sufficient water quality. However, various authors claim that the quantity of water is even more important than its quality in preventing the incidence of water-borne diseases (Konradsen *et al.*, 1997; Van der Hoek *et al.*, 1999), because an increased water availability is likely to improve personal hygiene. Vector-borne diseases are related to the condition of the river ecosystem, since many vectors, such as mosquitoes (malaria, dengue) and snails (schistosomiasis), need a certain condition of the river ecosystem as habitat during certain stages of their life-cycle (Jobin, 1999). The most direct impacts of river flows on human well-being are possibly floods and droughts: floods either take away waste, or bring dirt from upstream areas, and may also result in loss of human life. Droughts can create a hot, dry and dusty climate, with the risk of sandstorms, and associated negative health impacts.

Perception & experience

Perception and experience refer to the sense of belonging people may feel with respect to a certain area, the way an area provides the scene for traditional, cultural or religious ceremonies, or simply for recreation to relax and enjoy. Floods and droughts were mentioned as possible health threats. When floods and droughts do not pose a risk for human health, they may still be perceived as inconveniences, and negatively affect human well-being. Traditions and a sense of belongingness may be important components of peoples' identity. In many poor communities, income, food and health may be the top priority, but in traditional rural areas religion, and associated sacred or ceremonial sites, may play a much more profound role in peoples' life than in developed urban areas.

Second order components

Second order components include parts of well-being which have a more psychological connotation, such as reduced independence or self-esteem, or disruption of the social structure and cultural values. To understand the concept of mental well-being, it is useful to consider the type of impacts identified in Social Impact Assessments (SIA). Vanclay (2000) describes social impacts, besides changes in income and health, as changes in:

- peoples' way of life – how they live, work, play and interact with one another on a day-to-day basis;
- their culture – shared beliefs, customs, values and language or dialect;
- their community – its cohesion, stability, character, services and facilities;
- their fears and aspirations – their perceptions about their safety, their fears about the future of their community, and their aspirations for their future and the future of their children.

Because of these changes in peoples' way of living and social relationships, peoples' well-being may be negatively affected. Changes in these values can feedback and enhance the changes in income & food, health and perception & experience. Based on above types of changes, this thesis distinguishes second order components at three levels: 1) individual, psychological impacts for persons, 2) within the community, and 3) at the level between communities or between communities and authorities.

Summarising human well-being

This section summarises the identified components of well-being, with links between the river ecosystem and the components. For the second order components there are no direct links with the river ecosystem, but different levels of this component are listed.

First order components
Income & food. The river ecosystem contributes to this component through:

- possibilities for income-generating activities;
- provision of food, comprising the three main nutrient groups of starch, proteins and vitamins; and
- provision of 'free' materials for various purposes.

Health. The river ecosystem contributes to this component through:

- availability of water of sufficient quality and quantity for drinking and sanitation;
- prevention of suitable habitats for disease-vectors;
- availability of medicinal natural vegetation;
- processing of waste; and
- (prevention of) floods and droughts and related phenomena (e.g. sandstorms) (local climate and living conditions).

Perception & experience. The river ecosystem contributes to this component through:

- providing the scene for ceremonies;
- providing people with a sense of belonging to an area;
- providing opportunities for recreation; and
- inconveniences, for example flooding of homesteads, difficult transportation.

Second order components can be identified at three levels:

- individual;
- within the community; and
- between communities, or between communities and authorities.

Strictly speaking, changes in first order components as a result of changes in second order or other first order components are second order impacts, yet these impacts remain impacts in first order components. Quantification in this thesis will focus on assessing the direct impacts on income, health and perception, but will discuss the likeliness of further impacts through interactions between income, health, and various psychological, social, and cultural aspects.

3.3.5 Extension of the conceptual model

The conceptual model of Figure 3.1 can now be extended with the main components of well-being, at two different levels: first order and second order, which together constitute total human well-being (Figure 3.2). The arrows in two directions are both causal chains: as a result of changes in second order components, the value for the first order components can change as well.

People in different cultures, or in different stages of their life, may have different understandings about how important each of these factors is to their well-being, but for all people these components will be relevant.

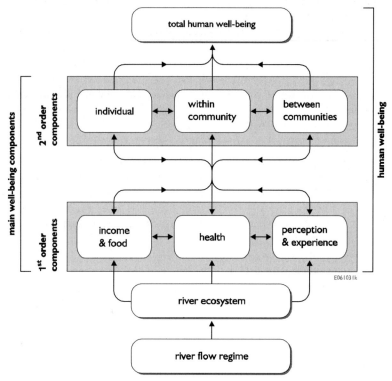

Figure 3.2 Extension of the conceptual model with components of human well-being

3.4 The river ecosystem and the link with human well-being

3.4.1 Use of the river ecosystem

In Chapter 2, the river ecosystem was defined as all components of the landscape that are directly linked to that river and all their life forms, including the source area, the channel from source to sea, riparian area, the water in the channel and its physical and chemical nature, associated groundwater in channel and bank areas, wetlands linked either through surface water or groundwater, floodplains, the estuary, and the near-shore marine ecosystem if this is clearly dependent on freshwater inputs. Of the various links between ecosystems and people (see Chapter 1), this thesis focuses on the use people make of the ecosystem. Other terms found in literature, related to use, are functions of ecosystems, ecosystem goods and services and values of ecosystems. De Groot (1992a) defines functions as "the capacity of natural processes and components to provide goods and services that satisfy human needs (directly and/or indirectly)". In this definition, functions are strongly related to the functioning of the ecosystem, or the ecosystem condition. Goods are normally understood as actual products obtained from nature, while services refer to all other benefits nature provides for people. The term 'uses' is found as referring to uses of goods and services, such as harvesting, but also as an encompassing term for goods and services (e.g. Turner et al., 2000). Values are often associated with economic valuation of goods and services (e.g. Barbier, 1997), but are also used in a similar meaning as functions or goods and services (e.g. Dugan, 1990).

Functions, goods and services and ecosystem use can be considered as three items in a chain, in which for example a function of the ecosystem is to provide fish habitat, leading to the availability of fish as part of goods and services, which people make use of through fishing. In this thesis however, these three terms are considered as one component of the conceptual model, referred to as the ecosystem goods and services component. A distinction is believed not to provide additional information for the purpose of this thesis. Functions, goods and services and use are together considered as the link between ecosystem condition and human well-being values. The term value is considered distinct from functions, goods and services and use, and refers to the contribution of the ecosystem to human well-being through income & food, health, and perception & experience, as well as the social, cultural and psychological values which are related to the second order components. Summarising, for the purpose of this thesis the river ecosystems is split into two components: ecosystem goods and services and ecosystem condition, which are both discussed in more detail in the following sections.

3.4.2 Ecosystem goods and services

Potentially, river ecosystems may have numerous goods and services. Various authors have attempted to identify and classify these goods and services or the functions (e.g. (Marchand & Toornstra, 1986; Dugan, 1990; ESCAP, 1992). Classifying is not an aim in itself, but a consistent classification prevents overlooking and double-counting of particular goods and services.

The classification adopted to describe goods and services in this thesis is the function classification by De Groot (1992b). His classification into seven categories: carrier, production, joint-production, habitat, signification, processing and regulation functions (Box 3.1) is an extension of one of the earliest classifications by Van der Maarel and Dauvelier (1978). De Groot's classification is used in this thesis as a checklist for potential ecosystem functions. Goods generally result from the natural and joint production functions, while all other functions provide services.

Of the seven categories of this classification, the habitat functions are considered superfluous for the purpose of this thesis. Habitat functions are actually part of the first four categories, as De Groot admits. Because the intrinsic value of nature is often overlooked, De Groot chose to add this separate category. Since in this thesis the focus is on the anthropocentric functions of the river ecosystem, omitting the habitat functions category provides a complete list of all potential functions of the river ecosystem for people.

The last two categories: processing functions and regulation functions can be considered conditions for the other five categories to take place at all. In such a case, regarding these functions as separate functions would result in double-counting. However, it is possible that the processing and regulation functions are benefiting ecosystems outside the ecosystem under consideration; in that case there is no double counting. Hence, processing and regulation functions can only be counted insofar as they are not already included as underlying the other five (De Groot, 1992b).

Box 3.1 De Groot's description of functions of ecosystems (De Groot, 1992b, p232,233)

> **Carrying functions** are characterised by the environment providing nothing more than space, substrate or backdrop for human activities.
> **Joint production functions** are defined by the types of relationships in which human decisions and inputs remain a dominant factor, but in which the environment is also actively involved, providing, for instance, soil fertility and the will-to-develop inherent in plants and animals.
> **Natural production functions** are characterised by the fact that the environment now produces (or has produced in history) largely on its own; human beings are only the harvesters. Harvesting, in this function category, is confined to physical entities (oil, wildlife etc.).
> **Signification functions** are defined by the fact that the environment again largely 'produces' on its own and human beings are only the 'harvesters', but 'harvesting' now lies in the cognitive and spiritual realms, e.g. those of science, cultural orientation and spiritual participation. The term 'signification' has been chosen as a reference to both the relatively superficial concept of 'to signal' and the deeper concept of 'to signify'.
> **Habitat functions** are those of which not humans, but the other intrinsically valuable inhabitants of the earth are the prime beneficiaries; habitat function is the provision of their ecological home.
> **Processing functions** are characterised by all the relationships in which people benefit from the capacity of the environment to undo the harm or risk inherent in human actions. In many of these functions (e.g. dilution, sequestration), the environment is relatively passive. In others (e.g. chemical transformation or the processing of organic waste), the environment plays a more active role.
> **Regulation functions** refer to the capacity of components of the environment to dampen harmful influences from other components. Often, this takes the form of a shield against too high levels of something, e.g. cosmic radiation or floods. In other instances, it is the dampening of processes that tend to go too fast or fluctuate too widely, e.g. soil erosion, the development of pests or river flow fluctuations.

It is important to realise that ecosystem functions have *the potential* to provide goods and services. If components of the ecosystem are not used, they should not be called goods. Goods and services lists of ecosystems contain potential goods and services. Certain services cannot always easily exist next to each other. For example, a wetland cannot at the same time be full of water for human use and accommodate floods. Another example is nature development on floodplains: while floodplain vegetation is appreciated for its beauty and habitat options, it may augment the risk of flooding, due to an increase of the roughness of the floodplain.

Additional aspects are sometimes included in other classifications, like *who uses* (ESCAP, 1992: subsistence and commercial use of the same resource), or *where use takes place* (De Groot, 1992b: internal and external use). Who uses and where use takes place are additions to ensure that the less obvious uses are not overlooked. Because in this thesis the focus is on the various groups of people using the river ecosystem, these aspects are automatically included and do not need to be part of the function classification.

The classification of functions with the list of the various goods and services involved allows for a further elaboration of the three identified first order well-being components. Table 3.1 lists what goods and services potentially contribute to the main and sub-components of well-being.

Table 3.1 Components of well-being, goods and services and function classes

Component of well-being	Sub-component	Possible goods and services	Function classes
Income & food	Income from various activities	Various goods and services	Natural production Joint production Signification function (for others, through ecotourism/ ecosystem protection)
Income & food	Food	Edible plants and animals	Natural production Joint production
Income & food	Free materials	Vegetation types providing materials	Natural production function Joint production function
Health	Drinking water	Water	Natural production Processing functions
Health	Hygiene (personal and environmental)	Water Dilution, degradation of pollution /flushing of waste	Processing functions
Health	Vector-borne diseases	Avoiding suitable conditions for disease vectors	Processing functions
Health	Health risks due to floods and droughts	Prevention of sandstorms, presence of vegetation	Regulation functions
Perception & experience	Religious/cultural/ traditional experience	Combination of goods & services (ecosystem integrity)	Signification functions Carrying functions
Perception & experience	Sense of belonging		Signification functions
Perception & experience	Recreation/enjoyment of nature		Signification functions Carrying functions
Perception & experience	Inconveniences due to floods and droughts	Prevention of sandstorms, presence of vegetation	Regulation functions
Perception & experience	Communication with outside world/transport		Carrying functions (travel over water)

3.4.3 Ecosystem condition

The condition of the ecosystem is the result of various interacting processes. An ecosystem can be considered as a network where various processes and species interact. Turner *et al.* (2000) define the wetland functioning as consisting of characteristics (e.g. substrate, species, water depth), structure (including communities of animals and vegetation) and processes. Changing one of these will influence all others. Some people may regard humans and their activities as part of this system, for example when harvesting of resources leads to cyclic rejuvenation, or when droppings of cattle provide necessary nutrients for vegetation. As was mentioned in Chapter 1, this influence of human use on the ecosystem is not considered in this thesis.

Chapter 3

3.4.4 Extension of the conceptual model

With the distinction of ecosystem goods and services and ecosystem condition, the conceptual model can be further extended (Figure 3.3). To keep the total figure orderly, the well-being components discussed in the previous section will further be represented as a single box referred to as main well-being components.

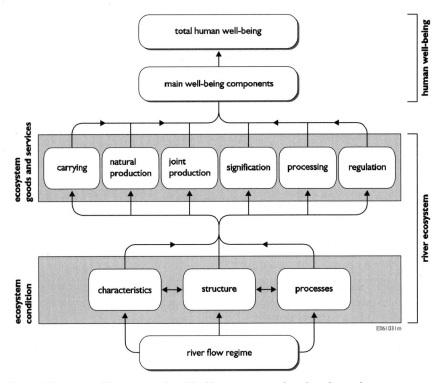

Figure 3.3 Extension of the conceptual model with ecosystem goods and services and ecosystem condition

3.5 The river flow regime and the link with the river ecosystem

3.5.1 Introduction

The link between the river flow regime and the river ecosystem is the main topic of investigation in the field of environmental flows. For the purpose of water resources management the flow regime is relevant in two locations:

- at the site of the ecosystem under consideration; and
- at the site where the flow regime can be managed, for example through reducing abstractions for off-stream use or through making releases from reservoirs.

The flow regime at the first location is actually part of the ecosystem under consideration and is therefore referred to as 'local' flow regime, whereas the second is referred to as the 'upstream' flow regime. Local and upstream flow regimes are linked through hydraulic laws. While for the local flow regime various hydrological parameters are relevant, such as inundation extent and depth, for the upstream flow regime it is mainly time series of discharge requirements which are needed.

3.5.2 Upstream flow regime

The upstream flow regime follows from natural inflows into the catchment and from various infrastructural interventions and related use and discharge of water. Assessing the upstream flow regime and its link with the local flow regime will typically require a combination of rainfall-runoff models and either hydrodynamic, routing or water balance models depending on the characteristics of the system under consideration.

3.5.3 Local flow regime: various components

As mentioned, the local flow regime is part of the ecosystem under consideration. The location for which the local flow regime needs to be assessed may be a single location, for example a wetland, but may also consist of various locations, such as various cross-sections along of a river.

Various authors have identified the natural dynamic character of the flow regime as important feature determining the type of ecosystem (Poff et al., 1997; Bunn & Arthington, 2002). In the previous section, the ecosystem was described as consisting of characteristics, structure and processes. The river flow regime can be considered one of these characteristics. In the concept of environmental flows, the river flow regime is considered to be *the* characteristic determining the condition of the ecosystem. For this reason, the flow regime is a separate component of the conceptual model. The presence of various factors influencing the condition of the ecosystem makes it sometimes hard to identify the exact role of the flow regime in this (Bunn & Arthington, 2002). The importance of other factors for the condition of the ecosystem should therefore not be ignored: these are part of the context, which will be discussed in Section 3.6.

Poff et al. (1997) identify five characteristics of the flow regime which they believe regulate ecological processes in river ecosystems: the magnitude, frequency, duration, timing, and rate of change of hydrological conditions. All characteristics show variations at time scales of hours, days, seasons, years and longer. To describe the flow regime various authors have developed indicators, reviewed by Olden and Poff (2003), of which the 32 indicators of hydrologic alteration (IHA) by Richter (1997) are perhaps the most commonly used. These indicators are part of the Range of Variability approach, one of the hydrological type environmental flow assessment methods. In this method the values of the indicators are not linked directly to specific conditions of the ecosystem.

In the DRIFT environmental flow assessment method, various components of the natural flow regime need to be identified, which can be described in terms of the characteristics of Poff *et al.* (1997). In an application of DRIFT to assess environmental flow requirements as part of the impact studies for the Lesotho Highland Water Project in Southern Africa, ten flow components were identified (Arthington *et al.*, 2003):

- dry season low flows;
- wet season low flows;
- intra-annual flood (magnitude) class I;
- intra-annual flood class II;
- intra-annual flood class III;
- intra-annual flood class IV;
- once in 2 year floods;
- once in 5 year flood;
- once in 10 year flood;
- once in 20 year flood.

For the natural flow regime the magnitude (discharge) for each of these components was determined, and for the inter-annual floods also the number of events per year. Based on this description of the natural flow regime, the deviation from the natural flow regime for a number of alternatives for dam construction and operation was derived. Subsequently, for various components of the ecosystem, for example various species of fish, the impacts were estimated for each combination of deviations from the natural flow regime (Arthington *et al.*, 2003).

Flow regimes for rivers vary across the world, and no generally applicable list of flow components and their characteristics can be given. In any environmental flow study, the natural flow regime needs to be analysed by hydrologists and ecologists together, in order to identify the flow components relevant for the ecosystem under consideration as well as the characteristics of these flow components in the natural (or before intervention) situation.

3.5.4 Extension of the conceptual model

Based on above discussion, the conceptual model is extended by dividing the flow regime block by a block representing the local flow regime and a block representing the upstream flow regime. The local flow regime is further divided into various flow characteristics, based on the ecosystem under consideration. As an example of possible flow characteristics, inter-annual and intra-annual characteristics are mentioned in the model, both consisting of flows of various magnitude, duration, extent, frequency and rate of change (Figure 3.4). For clarity reasons, the ecosystem part is further represented by two blocks: ecosystem condition and ecosystem goods and services.

With this extension a detailed understanding of the links between flow regime and human well-being is derived. However, to be able to understand not just the link with, but also the importance of the flow regime for human well-being, further extensions are required. This is the topic of the next section.

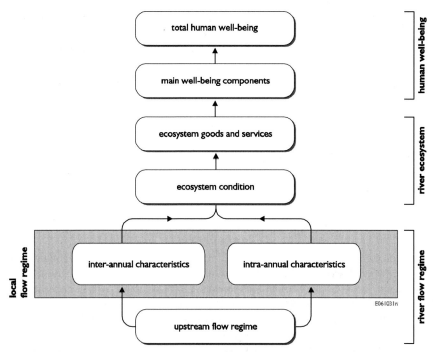

Figure 3.4 Extension of the conceptual model with local and upstream flow regimes

3.6 The context: understanding the importance of river flows and related ecosystems for people

3.6.1 Context

Knowing what goods and services of the river ecosystem are used is not sufficient to understand the importance of the river ecosystem and the river flow regime for people.

The question of how important something is can only be answered when the relationships between a river's flow regime and human well-being are understood in the light of the full context of all components. For example, when people have additional sources of income next to the income they obtain from the river ecosystem, the river ecosystem is of less importance to these people than to the people who obtain their entire income from the river ecosystem. The context is here defined as all factors that affect the relationships between flow regime, river ecosystem and human well-being.

Taking the context into account is useful for the following reasons:

- Considering the context helps to assess the impacts of ecosystem and flow regime changes and the importance of these impacts.
- Understanding the importance of environmental flows for different groups of people will be the starting point for compensation and mitigation.
- Assessing the context may reveal opportunities for measures which are either more effective or more efficient, or which are prerequisites for environmental flows to have the desired impact of the ecosystem. This prevents a wastage of water.

In certain situations, considering the context may reveal that the ecosystem is of utmost importance to the well-being of the people, or, to the contrary, that it is not that important. Both outcomes will contribute to better informed decision-making in water resources management.

The context may play a role at all links of the conceptual model:

- Link between total human well-being and the main well-being components: general welfare conditions and cultural aspects.
- Link between well-being components and ecosystem goods and services: availability of alternatives.
- Link between ecosystem goods and services and ecosystem condition: access to ecosystem goods and services.
- Link between ecosystem condition and local flow regime: other factors influencing the condition of the ecosystem.
- Link between upstream and local flow regime: local infrastructure and abstractions.

An additional aspect of the context, which is not related to a particular link in the conceptual model, is the possible negative impact of natural flow regimes on people. Each aspect of the context will be discussed in more detail in the following sections.

General welfare conditions and cultural aspects

Differences in cultural and development context may mean that certain aspects of well-being are perceived to be more important than others. According to Maslow's hierarchy of needs (Maslow, 1954), certain needs only become apparent when lower level needs are fulfilled. In poor communities, food and income may have the highest priority, while in more developed countries recreation may play an important role. This part of the context determines the relative importance of the main well-being components, and since some well-being components may have stronger links with the river ecosystem and river flow regime than others, it determines the importance of the river flow regime.

Availability of alternatives

Ecosystem goods and services of the particular river or wetland under consideration will generally not be the only contributors to the main well-being components. Water may be available from other sources, or people may obtain part of their income from not-ecosystem related sources. If there are alternatives to the benefits of environmental flows, the importance of environmental flows for human well-being is reduced. Alternatives can be divided in three types:

1. Current use of goods and services provided by environmental flows is only part of the total contribution to well-being. For example, a household receives 50% of its income through fisheries and 50% through a car rental business. Generally, poor rural households diversify their livelihoods as a survival strategy (Narayan, 1999).
2. In the current situation alternatives are not used, but in case of emergency those are available: for example less preferred income-generating activities, support from family members.
3. In the current situation alternatives are not available, but measures could be taken to provide these, which may be more cost-efficient than maintaining or increasing river flows. For example, providing piped water supply instead of maintaining river flows for domestic purposes.

Access to ecosystem goods and services

When people do not have access to goods and services, a healthy ecosystem will not contribute to their well-being. An example is the situation in which certain river and floodplain areas are leased or possessed by rich and powerful people and out of reach for the poor traditional fishermen. Legal and regulatory changes are required at first to let the fishermen benefit from the ecosystem condition.

Other factors influencing the condition of the ecosystem

An ecosystem is to be understood as a complex network of structures, characteristics and components, as discussed before. The river flow regime is but one factor in this complex network. Changes in other parts of this network, as a result of over-exploitation or pollution will change the ecosystem condition, meaning that established relationships between flow regime and ecosystem condition are no longer valid.

Local infrastructure and abstractions

When the upstream flow regime is to be maintained to obtain the required local flow regime, this does not yet guarantee that the required local flow regime is achieved. Local infrastructure can influence the hydraulic relationship between upstream and local flow regimes. An example is the closing of connections between river and floodplains by floodplain farmers in Bangladesh. A connection to the floodplain is required for river fish to spawn and graze on the floodplain. In such a case, maintenance of river flows will not lead to an increased fish population without the co-operation of the local farmers.

Negative impacts of natural flow regimes

From an ecosystem point of view natural flows should be maintained as much as possible. Protection of the valued features of the river ecosystem focuses on the benign impacts of river flows. However, natural river flows may have negative effects as well: floods and droughts. Important as these natural characteristics may be for sustaining the natural ecosystem, they may be disastrous for people. This part of the context should follow from the assessment of the importance of river flows and the

Chapter 3

river ecosystem for human well-being. Instead of only defining a minimally required flow characteristic it would be better to define a range with a maximum, or at least an indication of whether the water level or flow velocity required for a certain function has negative effects on other functions as well. Although the negative effects of river flows are not explicitly considered in the conceptual model, they form a constraint to the local flow regime. In systems were floods and droughts are managed as part of the implementation of environmental flows, human use can be protected through careful timing and planning of both floods and droughts or through actual protection works.

3.6.2 Extension of the conceptual model

The conceptual model presented in Figure 3.4, but with a simplified presentation of the local flow regime, can now be extended with the context (Figure 3.5). The negative impacts of the natural flow regime are not linked to a specific part of the conceptual model, but form a constraint to the assessed environmental flows. Therefore, this aspect of the context is not included in the conceptual model.

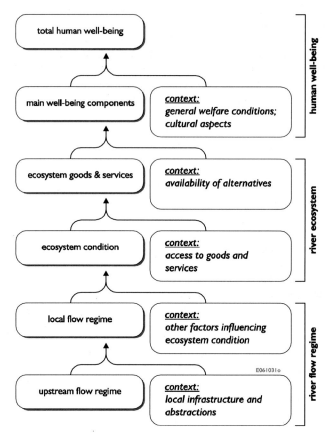

Figure 3.5 Extension of the conceptual model with the context

The purpose of the conceptual model is to provide an overview of all factors influencing peoples' well-being. How the conceptual model can be used to quantify social equity and how it connects to IWRM and environmental flow assessment methods is reflected upon in the next section.

3.7 Applying the conceptual model as part of IWRM analysis: a stepwise approach

The previous sections built a complete conceptual model. This section discusses how the conceptual model fits in the systems analysis approach of IWRM, and how social equity can be quantified through applying the conceptual model. This requires at first a further elaboration of social equity. Subsequently, the conceptual model is translated in steps for a practical assessment. These steps are then compared to the steps of the systems analysis approach.

3.7.1 Social equity: the importance of river flows and the river ecosystem for different groups of people

The concept of social equity

Social equity is one of the three criteria IWRM should comply with, the other two being economic efficiency and environmental sustainability. It refers to the distribution of benefits and losses over different groups of people as the result of development projects or strategies applied to achieve better water resources management. So, human well-being should not be assessed as one total value for a certain area or as a result of a certain measure, but rather the increase or decrease in human well-being for different groups of people should be assessed.

Differences between groups of people can have two causes: first, the link between peoples' well-being and the ecosystem may be different. A fisherman for example, depends on different flow regime characteristics than a recession farmer or pastoralist. Second, the context for different people may be different: between fishermen, there may be differences with respect to vulnerability or power relationships, meaning for example a difference in availability of alternatives.

Stakeholder identification

To assess social equity, it is at first important to identify the relevant stakeholder groups. Stakeholders are defined as all people whose well-being will be affected by changes in the river ecosystem or river flow regime. A group can be identified as a separate group when it is likely to experience different effects from changes in the river ecosystem and river flow regime. Effects can be different because they affect different aspects of somebody's well-being or because the effect is of a different severity. From the analysis of links and context it will be possible to identify groups, but the following characteristics may be useful to distinguish groups in an early stage:

- type of income generation activities;
- poverty level;
- age;
- gender;
- location along a river.

Type of income-generating activities
Different activities may require different river flows or depend to a smaller or larger extent on the river ecosystem. Often distinguished are communities of fishermen, pastoralists and (recession) farmers. Other activities can be thought of, such as reed harvesting and bird-hunting.

Poverty level
Poverty level influences the effects on people in two main ways:

- poor people are more vulnerable to changes, because they have less alternatives and less reserves;
- in poor communities income and food are probably receiving the highest priority, other aspects of well-being, such as recreation may be considered less important, as was mentioned in section 3.6 as part of the cultural and development context.

Chambers (1983; 1995) identified eight dimensions of deprivation:

1. poverty: lack of physical necessities, assets and income;
2. social inferiority: being part of a 'lower' social group as a result of gender, caste or ethnicity;
3. isolation: being peripheral and cut off from communication, contacts and information;
4. physical weakness: especially in rural areas, a healthy body is a major resource;
5. vulnerability: exposure to stress and shocks, and a lack of means to cope without damaging loss (defencelessness);
6. seasonality: well-being in rural areas often varies with the season;
7. powerlessness: poor household have limited knowledge of law and no access to legal advice, they are easily exploited because they have not much choice than to take debt against high interest rated, or to work for low wages;
8. humiliation: self-respect, freedom from independence.

The World Bank (2000, in GWP (2003)) found that "while poverty is often described in socio-economic terms, focusing on inadequateness of meeting the basic needs of household members, people who endure poverty may define their condition differently, seeing their problems as characterised less by economic disadvantage or lack of service access than by powerlessness, voicelessness, insecurity and fear – socio-political factors". These various dimensions of poverty should be recognised to identify stakeholder groups which are likely to experience different impacts from flow regime changes.

Age
People in different age groups can experience different effects. This can be the result of the fact that different age groups are responsible for different activities. For example, children may be responsible for herding cows or fetching water. On the other hand, if large structural changes are occurring in a community, old people may

be less flexible to change. Their well-being can be stronger affected than young peoples' well-being, although the changes in their environment are the same.

Gender
Because women are often appointed specific tasks, the effect on men and women can be different. If the income-generating activities carried out by women are affected, the women may become more dependent on their husband or family-in-law.

Location along a river
Changes in the river flow regime may have different effects along a river. In downstream areas for example, salt water intrusion may become a problem. Upstream users are normally in an advantaged position. However, when confluences take place downstream of interventions, along the regulated river the downstream users are in a better position than the upstream users. Also the type and availability of goods and services change along the river, which influences which people can use what.

Links between distinguishing characteristics
Some of these characteristics to distinguish groups of people may be related. For example, the poorest people often involve in similar type of work, like fisheries, and live on or near the river banks. The discussion of the characteristics to distinguish groups of people mainly has the purpose of indicating possible ways through which the well-being of various groups of people can be affected.

3.7.2 Steps to take as part of IWRM analysis

To enable application of the conceptual model as part of an IWRM analysis, five steps for applying the conceptual model are defined:

1. identifying stakeholder groups;
2. assessing the relationship between human well-being and the river ecosystem;
3. assessing the relationship between the river ecosystem and the local flow regime;
4. assessing the relationship between the local flow regime and the upstream flow regime; and
5. estimating impacts of water resources management measures on the well-being of the identified stakeholder groups.

Steps 2, 3 and 4 relate directly to parts of the conceptual model as presented in Figure 3.6. Steps 1 and 5 deal with further application of the assessed relationships. In step 5 the changes in the upstream flow regime, resulting from water resources management measures, are translated into impacts on human well-being for each of the, in step 1, identified stakeholder groups. This translation is done with the relationships assessed in steps 2, 3 and 4.

Step 1: Identifying stakeholder groups

To quantify social equity it is important that groups of stakeholders which are expected to experience different impacts from flow regime changes are distinguished at an early stage. Groups can be distinguished based both on the link they have with the ecosystem and the flow regime and on differences in context. Social equity does not require a further extension of the conceptual model; the model should be applied for each of the identified groups. The lower part of the model, the links between flow regime and ecosystem condition will generally not lead to differences between groups of people. Differences will follow from the relationships between ecosystem condition, availability of goods and services and human well-being. Identification of stakeholders includes an assessment of how many households or people each group counts, as well as information on where to find them, or representatives of the group.

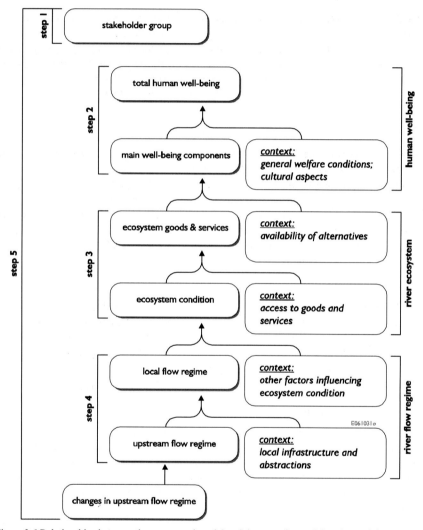

Figure 3.6 Relationships between the conceptual model and the steps for applying the model

Step 2: Assessing the relationship between human well-being and the river ecosystem

This step requires the identification of indicators for human well-being linked to each of the main well-being components and tailored to the situation. For each group of people a quantified relationship needs to be established. Information on potential ecosystem goods and services as well as on the ecosystem condition and local flow regime over the past years will facilitate the communication with the local communities and the analysis in this step.

Step 3: Assessing the relationship between the river ecosystem and the local flow regime

The link between the biophysical ecosystem and various flow characteristics is the main focus of the existing environmental flow methods. Various methods may be appropriate to apply in this step, but it is important that the parts of the ecosystem important for the local population can be included in the selected environmental flow assessment method. Therefore, holistic, scenario-based methods will be the most appropriate.

Step 4: Assessing the relationship between the local flow regime and the upstream flow regime

The often applied hydraulic and/or water balance and allocation models take care of what is required in this step: calculating how much water enters a river basin in a certain time period, how much is used by the various water demand sectors, and subsequently how much is left in the river itself. Processing of the resulting time series at various parts of the river basin results in the required characteristics of the local flow regime.

Step 5: Estimating impacts of water resources management measures on the well-being of the identified stakeholder groups

With the established relationships of steps 2, 3 and 4, the well-being impacts of alternative water resources management strategies can be assessed. This should be done for the various groups of stakeholders of step 1. The impacts on the well-being of the various groups of people for all alternative strategies should be communicated to the water resources management decision-makers.

3.7.3 Application in the system analysis approach

To ensure that the conceptual model can be applied in IWRM, the conceptual model and the five steps should fit into the systems analysis approach as described in Chapter 2. Step 1, identifying stakeholders, is similar to the first step in the systems analysis approach. Stakeholder analysis in IWRM will focus on government authorities, NGOs, commercial organisations and other representatives of the civil society. Such an analysis is often referred to as actor analysis, in which attention is paid not only to who the stakeholders are, but also to their objectives, perceptions and resources, as well as to the relationships with other actors (Hermans, 2005). The stakeholder identification for the application of the conceptual model will focus on specific stakeholder groups, for which certain information is required in more detail. The actor analysis of IWRM can provide information on, among other things, the various groups of users of river ecosystems, the size of these group, and their problems.

In the systems analysis approach it is analysed to what extent a water resources system contributes to the objectives defined for it. The relationship between river flows, the river ecosystem and human well-being discussed in this chapter are an elaboration of a part of the water resources system. The conceptual model therefore fits into the systems analysis approach through providing part of the model of the water resources system.

This model is required at first to assess whether current water resources management may lead to problems either at present or in the future. If so, measures will need to be developed and impacts of these measures estimated. With the part of the model resulting from steps 2, 3 and 4 the consequences of water resources management measures and strategies on the river ecosystem and subsequently on the people making use of this ecosystem can be predicted. This evaluation of impacts of step 5 is again part of the total evaluation of impacts.

The links between the conceptual model and the systems analysis approach presented in Figure 2.4 are shown in Figure 3.7. This figure shows that steps 2, 3 and 4 contribute to a better understanding (and modelling) of the water resources system.

Impacts of water resources management strategies need to be expressed in terms of objective-related criteria, which will be different in each situation. If the concept of IWRM is applied well, enhancing social equity should be expressed as objective for water resources management. Indicators for the first and second order human well-being values can be linked to the criteria identified for evaluation of the impacts of the alternative water resources management strategies.

The conceptual model and steps help to understand the water resources system and to predict impact of water resources measures, but the results of the application of the conceptual model can also provide starting-points for water resources management measures. It was mentioned before that the analysis of the context could help to identify measures which are either more effective or more efficient to improve well-being than adjusting flow regimes. Below, some measures will be discussed for each level of the conceptual model starting at the upstream flow regime moving upwards to total human well-being.

Measures to influence the upstream flow regime
The most direct way to sustain the river ecosystem, and the basic idea behind environmental flow assessments, is the maintenance of a near natural upstream flow regime. The upstream flow regime can be changed through reduced abstractions for other water users, or through the releases of flows for ecosystem water use.

Measures to influence the local flow regime
When there is not sufficient water to maintain a complete ecosystem, changing the dimensions of the river ecosystem may result in obtaining the desired local flow regime in a limited area of the river ecosystem. The part of the ecosystem subject to a natural flow regime may be in a good condition, although for specific species a minimum habitat area should be observed. A smaller ecosystem will change the availability of goods and services. In situations where large communities used to make use of the larger ecosystem, this may result in a reduced level of well-being and in over-exploitation of the system.

Measures to influence the ecosystem condition
River flows are but one factor sustaining the river ecosystem. Pollution and over-exploitation may constitute additional threats next to changed flow regimes. Pollution can be prevented through various ways: raising awareness, collecting and purifying waste water, or diverting pollution flows to less ecologically valuable channels or lakes. Ecosystem protection requires measures of a legal manner: providing and maintaining legislation, as well a raising public awareness. Reports of the International Union of the Conservation of Nature (IUCN) provide information on how ecosystem resources can be protected through economic incentives (McNeely, 1988) and management taking into account community interests (Pirot *et al.*, 2000) .

Measures to influence the availability of goods and services
When access is not legally restricted in order to protect natural areas, social and cultural factors may still limit access to the ecosystem for particular groups of people. To protect the less powerful people, authorities can issue licenses for use of the resources, or take measures against illegal occupancy of ecosystems by powerful groups. This requires legislation and law enforcement. In countries where corruption is prevailing, rich people are often in an advantaged position.

Measures to influence the various values of human well-being
The first measure to protect human well-being related to ecosystem goods and services is the protection of the ecosystem. However, one can also think of providing alternatives for ecosystem use. Water can be supplied from different sources, such as from reservoirs, groundwater or through rainwater harvesting, and food and other resources can be obtained from other areas. Since the alternative sources of water may be taken from other parts of the same water resources system, it should be analysed whether the shift to other sources is indeed desirable. Economic development and education may provide alternative employment opportunities, with possibly equal impacts on human well-being. Other than providing aid in the form of food and money, the provision of alternative jobs will preserve peoples' independence and social structure.

Chapter 3

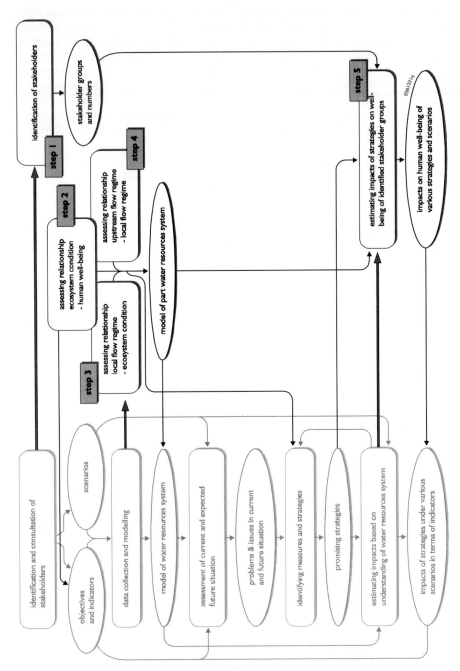

Figure 3.7 Links between the conceptual model and systems analysis

3.8 Conclusions

To include human well-being related to environmental flows in IWRM, Chapter 2 identified that there is a need to elaborate the social equity criterion to represent full human well-being, and to derive quantifiable links with the river ecosystem. With such links and indicators, the impacts of water resources management strategies on social equity, via river ecosystems, can be quantified. In this chapter, a conceptual model has been constructed consisting not only of the links between human well-being, the river ecosystem and the river flow regime, but also of the context in which to consider these links. Five steps were identified to apply the conceptual model as part of the systems analysis in IWRM.

Links between human well-being, the river ecosystem and the river flow regime

To enable quantification of the relationships, a thorough conceptual understanding of the relationships between river flows, the river ecosystem and human well-being is required. Through unravelling of the concepts of the river flow regime, the river ecosystem and human well-being, and the relationships between these concepts, this chapter built a conceptual model. Although the focus of this thesis is on the relationship between human well-being and the river ecosystem, it is required to take into account the other relationships as well. The three linked relationships are needed to assess changes in human well-being resulting from changes in the flow regime.

To describe human well-being, various components of well-being were distinguished. Three components were identified which are directly related to the river ecosystem: income & food, health and perception & experience. The socio-economic system is complex and dynamic and changes in these three components of well-being are likely to impact other components of well-being such as independence, social structure and other psycho-social factors. These components of well-being are in this thesis referred to as second order, because they are affected indirectly by the changes in the river ecosystem. The components directly related to the ecosystem are referred to as first order. As a result of changes in the second order components, the first order components can change further, either in a positive or a negative direction.

Context

An important component of the conceptual model is the context. The context pays attention to factors that can influence the identified links between the river flow regime, the river ecosystem and human well-being. Only through giving due consideration to the context, the importance of the river ecosystem, and of river flows, for human well-being can be understood.

Application in IWRM

The purpose of the conceptual model is to aid analysis of impacts on social equity of alternative water resources management strategies. Indicators related to decision-making criteria can be defined in practical situations based on the identified components of well-being. With the links of the conceptual model, including the context, the impacts of various measures can be described in terms of these indicators.

To provide quantified insight in social equity, the links of the conceptual model need to be derived for various groups of people that are likely to experience different effects from changes in river flow regimes. Identifying the stakeholder groups is therefore an important step in the application of the conceptual model.

To apply the conceptual model a total of five steps are formulated, each of which has logical links with the components of the systems analysis approach:

1. identifying stakeholder groups;
2. assessing the relationship between human well-being and the river ecosystem;
3. assessing the relationship between the river ecosystem and the local flow regime (hydrological characteristics);
4. assessing the relationship between the local flow regime and the upstream flow regime; and
5. estimating impacts of water resources management measures on the well-being of the identified stakeholder groups.

The conceptual model builds upon knowledge generated through existing environmental flow assessment methods. These methods are required to derive in a structured way the relationship between the local flow regime and the river ecosystem in step 3 of this approach.

Overall conclusion

The proposed model will contribute to a better understanding of the relationships between human well-being, the river ecosystem and the river flow regime, and will facilitate the quantification of changes in the river flow regime or the river ecosystem in terms of human well-being and social equity. In the next two chapters the application of the conceptual model in Bangladesh (Chapter 4) and in Iran (Chapter 5) is described. A synthesis of these applications, leading to practical guidelines is given in Chapter 6.

4 Well-being values of the Surma and Kushiyara Rivers and floodplain, Bangladesh

4.1 Introduction

Bangladesh shares 54 of its many rivers with India. One of these rivers is the Barak, which upon entering Bangladesh, bifurcates into the Surma River and the Kushiyara River (Figure 4.1). Annual flooding of the area between the Surma River and the Kushiyara River has resulted in a fertile system with fisheries and agriculture, both of which are important for income and food for the inhabitants. Through planned upstream interventions in India, as well as through natural morphological changes, the flow regime of both the Surma River and the Kushiyara River will change. Therefore, it is necessary to assess the importance of the current flow regime for the well-being of the people living in the area. This assessment is carried out as a first application of the conceptual model as described in Chapter 3 and is discussed in this chapter. The case study served to test and further develop the conceptual model.

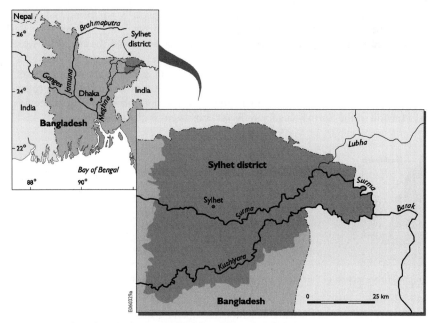

Figure 4.1 Location of Barak, Surma and Kushiyara Rivers in Bangladesh

The next section describes the characteristics of Bangladesh and in particular of the Surma-Kushiyara floodplain area as well as the framework in which the case study has been conducted. Section 4.3 discusses the general approach used in the case study and the methods used for data collection. The results are described in five sections: section 4.4 identifies stakeholder groups and numbers, while section 4.5 discusses for each of the stakeholder groups their link with the river ecosystem. The link between ecosystem goods and services and river flows is discussed in sections 4.6 and 4.7. Section 4.8 combines all links and discusses the likely impacts of three possible changes to the flow regime. Section 4.9 draws conclusions on three levels: 1) human well-being aspects of environmental flows for the Surma-Kushiyara Rivers and floodplain, 2) suitability of the methods used to assess environmental flows from a well-being perspective, and 3) the usefulness of the conceptual model for the main topic of this dissertation: assessing human well-being values of environmental flows.

4.2 Background & objective of the case study

4.2.1 Description of the area

The Surma and Kushiyara Rivers flow through the northeast region of Bangladesh. This region is characterised by its many wetlands, which are rich natural resource areas. Many of these wetland areas are part of the year isolated lakes, while at other times they are connected to the rivers. This link with the river is considered very important for sustenance of the natural ecosystem.

The river system of the Surma-Kushiyara basin is very complex. The area is confined by hills on both sides of the catchment. From these hills many small rivers flow down to join the Surma River or the Kushiyara River. Especially in spring, when high rainfall occurs, the quick run-off from the hill catchments leads to dangerous flash floods. During the pre-monsoon season, in April and May, and the monsoon season from June to September, river discharges increase and a large part of the area is flooded. The floodplains are bowl-shaped, resulting in some of the deepest parts in permanent water bodies, called beels. A floodplain area with various beels is referred to as haor. Through khals, small canals, the river and floodplain can be connected. A cross-section of the Surma-Kushiyara River ecosystem is shown in Figure 4.2.

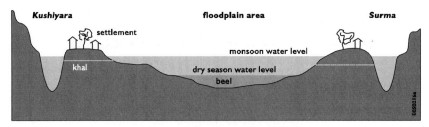

Figure 4.2 Cross section of Surma-Kushiyara River ecosystem in downstream direction (not to scale)

The eastern part of the Surma-Kushiyara floodplain, which is the area considered in this case study, is mainly part of the agro-ecological zone 20 (AEZ 20) (SLI & NHC, 1994c). This agro-ecological zone has high rainfall, ranging from 2,500 mm in the south to over 5,000 mm annually in the north. For the entire northeast region, it is estimated that 40% of the total surface water supply originates from rainfall within

Bangladesh, and 60% from rainfall in India (SLI & NHC, 1993b). Compared with the rest of Bangladesh, this area is highly prone to flash and monsoon flooding. Although flooding causes problems, the economic system depends on fisheries and recession agriculture which both benefit from flooding.

Agriculture and fisheries are the main income-generating activities in the Surma-Kushiyara basin which are related to land and water. What crops can be cultivated is highly determined by the elevation of the arable fields. An often used land-classification system in Bangladesh is a list proposed by the Master Plan Organisation (MPO)2, based on land elevation. Table 4.1 lists the land classes with flooding depth, the percentage of the land type classified as agro-ecological zone 20, and the main crops cultivated. Aus, aman and boro are the three main varieties of rice cultivated in the area. Aus is cultivated in the premonsoon period (April-May), aman in the post-monsoon period (August-November), and boro is a winter crop (December–April). Because of the variations in water level over the year, the three crops are cultivated at fields with different elevations. The highest elevated fields are used for aus, and the lowest, near the permanent water bodies, for boro. Fields with an elevation in between are cultivated with aman.

Table 4.1 MPO land classification with main crops

MPO class	Description	Flooding depth (cm)	% in AEZ 20	Main crops
F0	Highland	0 – 30	5	Local and HYV* aus Local and HYV transplanted aman Jute
F1	Medium highland	30 -90	25	Local and HYV aus Local and HYV transplanted aman HYV boro
F2	Medium lowland	90-180	20	Deepwater aman HYV boro Wheat
F3	Lowland	> 180	36	Local and HYV boro
F4	Very lowland	> 180 m, may increase > 6m depth and does not permit any crop to grow in monsoon	<1	Local boro
Homesteads/water			14	

* HYV stands for high yielding variety

Along the Surma and Kushiyara Rivers two types of flood protection can be found: submersible embankments and full flood protection, in addition there are areas

---

[2] The MPO was created in 1983 under the Ministry of Water Resources. In 1992 MPO was renamed as Water Resources Planning Organisation (WARPO) which still exists.

protected through natural levees only. Submersible embankments are meant to protect the area from flash floods. This is important to protect boro rice, which is harvested during the flash flood season. Monsoon flooding cannot be prevented by submersible embankments. Plans are made to construct full flood protection along the left bank of the Surma River and the right bank of the Kushiyara River, protecting the entire area between Surma and Kushiyara Rivers. Still, full flood protection may not be able to protect the area from seasonal flooding, due to a combination of high rainfall and drainage congestion downstream. No detailed information on the status of flood protection in the area is available.

Administratively, Bangladesh is divided into 64 districts or *zila's*. These districts are divided in sub-districts, in Bangla called *upazila's* or *thana's*. The sub-districts are split into unions, which at their turn comprise several villages. A large part of the Surma-Kushiyara basin lies within the boundaries of Sylhet district (see Figure 4.1). The total population of Sylhet district was slightly over 2 million in 1991 (BBS, 1996), with, at that time, a growth rate of 1.93%. Currently, the growth rate is decreasing. The total area of Sylhet district is 3,490 km2, which means a population density, in 1991, of 617 people/km2. The literacy rate of this district is 34%.

The distribution of the Barak River inflow over the Surma and the Kushiyara Rivers is changing as a result of morphological processes: the proportion of Barak discharge flowing into the Kushiyara River is increasing, while discharge into the Surma River is decreasing. In the future, the flow regime may change further when planned upstream developments in India will be completed. Approximately 200 km upstream of the India-Bangladesh border, the Tipaimukh dam is planned. Besides using the stored water for hydropower generation, there are plans to irrigate the Cachar Plain downstream of the dam. For this purpose a barrage will be build on the Barak at Fulerthal, 100 km upstream of the bifurcation point at Amalshid (SLI & NHC, 1993b). Flood control is mentioned as a third purpose of the reservoir (SLI & NHC, 1993a). These developments will largely influence the flow regime of the Barak River, and consequently of the Surma and the Kushiyara Rivers.

An India-Bangladesh Joint River Commission is in place to discuss the management of various transboundary rivers. Already in 1966, the work of this commission led to the signing of a treaty on Ganges water use. Nowadays, among other issues, the developments on the Barak river are on the agenda of the JRC. Discussion of this issue requires information on the impacts of changing the Barak flow regime.

4.2.2 Framework and objective of the case study

Understanding the instream flow requirements is highly relevant for a country at the downstream end of many transboundary rivers. Environmental flow requirements are understood as part of the total instream flow requirements. In the framework of a capacity building project, the Bangladesh University of Engineering and Technology (BUET) has engaged in an applied research project on 'Management of rivers for in-stream requirement and ecological protection' (Bari & Marchand, 2006). This research topic was selected in close co-operation with the authorities responsible for water resources management in Bangladesh. In this research, environmental flow assessment methods were identified, their suitability for application in Bangladesh investigated, and subsequently applied on three river systems. The Surma-Kushiyara system was one of the selected river systems, the other two being the Teesta in the dry northwest of Bangladesh and the Gorai, an off-take of the Ganges in the southwest. The research carried out focused on hydrological environmental flow assessment methods and on the ecological condition of the river. The case study described in this chapter was carried out as additional research on the well-being related to the river ecosystem, and was conducted in close co-operation with BUET.

The Surma-Kushiyara river system was selected because of the previously mentioned planned upstream developments in India, as well as ongoing morphological processes at the bifurcation point. These developments, which will change the natural flow regime, may threaten the current system of fisheries and paddy cultivation which depends on the pattern of annual flooding of the floodplain areas. An assessment of environmental flow requirements is required to support discussions with India on the effects of their development plans.

The objective for the case study discussed in this chapter is therefore:

> *To assess the importance of the river flow regime and related ecosystem condition for the well-being of people living in or near the Surma-Kushiyara river and floodplain ecosystem.*

4.3 Method: approach & data collection

4.3.1 General approach

In this case study the focus is on a qualitative assessment of the first order human well-being values related to the river ecosystem and the hydrological processes which play a role in sustaining the ecosystem goods and services. The general approach consists of the five steps identified in Chapter 3, translated to the situation of this case study:

1. identifying stakeholder groups;
2. assessing the relationship between human well-being and the Surma-Kushiyara river and floodplain ecosystem;
3. assessing the relationship between the Surma-Kushiyara river and floodplain ecosystem and the local flow regime;
4. assessing the relationship between the local flow regime and the upstream flow regime; and
5. estimating impacts of changes in upstream flow regime on the well-being of the identified stakeholder groups.

Step 1. Identifying stakeholders groups

Part of the Surma-Kushiyara floodplain area is selected for this case study. All people whose well-being will be affected by changes in the flow regime of the Surma or the Kushiyara are considered to be stakeholders. Because the selected area is almost entirely inundated during the monsoon season, all inhabitants of the floodplain area were identified as stakeholders. A first division of stakeholders was made based on a population census from the Bangladesh Bureau of Statistics. A census is conducted every 10 years and is published five years later. The most recent census data available were therefore the data for 1991 (BBS, 1996). Based on estimated growth rates

(WARPO, 2000), the figures for 1991 were updated for the year 2000 and predicted for the year 2025. The results of this step are discussed in section 4.4.

Step 2. Relationship between human well-being and the Surma-Kushiyara river and floodplain ecosystem

The relationship between human well-being and the river ecosystem is assessed through household interviews in four selected villages. Combined with the statistical information of step 1, this step resulted in an overview of the importance of the river ecosystem for distinct stakeholder groups in the selected floodplain area. The focus is on the direct use of river ecosystem goods and services, and the first order impacts of changes in the flow regime. Second order impacts are not considered in this case study. Data collection methods and site selection are discussed in sections 4.3.2 and 4.3.3 respectively. The results of this step are discussed in section 4.5.

Step 3. Relationship between the Surma-Kushiyara river and floodplain ecosystem and the local flow regime

In this step the hydrological processes which play a role in maintaining the main goods and services are assessed, based on data collected from the local population and information from consultancy reports on the Northeast Regional Water Management Project/Flood Action Plan (FAP) 6 (SLI & NHC, 1993b). The results of this step are discussed in section 4.6.

Step 4. Relationship between the local flow regime and the upstream flow regime

The relationship between the relevant local hydrological parameters and the upstream flow regime was assessed through post-processing of simulation results of a 1-D Mike-11 model. This Mike-11 model comprises the entire northeast Bangladesh region and was available at the Institute for Water Modelling of Bangladesh, one of the partners in the larger capacity building project. The results of this step are discussed in section 4.7.

Step 5. Impacts of changes in upstream flow regime on the well-being of the identified stakeholder groups

In this step, likely well-being consequences of changed flow regimes are assessed for the identified stakeholder groups, based on a combination of the links established in steps 2 to 4. The results of this step are discussed in section 4.8.

4.3.2 Primary data collection

Interviews at the household level

The main method for primary data collection used in this case study was interviews with people living in villages along the river. The interviews focussed on various well-being components and the role the river ecosystem plays in maintaining or improving the well-being level. The household level was chosen as a suitable level for carrying out the interviews, because this is the level at which food, income and other resources are shared. Four villages were selected along the Surma river (see next section). In total 62 interviews were used for the analysis. The number of interviews conducted in each village is shown in Table 4.2. The interviews were semi-structured and were conducted with the aid of an interpreter.

Table 4.2 Number of interviews conducted in each village

Village	Number of households	Conducted interviews
Bhabanipur	46	5
Khulurmati	101	20
Jalalnagar	119	16
Sachan	79	21

4.3.3 Selection of sites and participants

The environmental flows study for the Surma and the Kushiyara rivers aims at defining the flow regime required from India, in this case, through the Barak river. Because the northeast region of Bangladesh has high precipitation, in downstream direction along the Surma and the Kushiyara the proportion of the total discharge originating from the Barak river will gradually reduce as the run-off component from adjacent land increases. For this reason, the focus in this case study has been on the people living directly downstream of the India-Bangladesh border. The floodplain area selected for the case study is located between the Surma and the Kushiyara river. Of the total 11 *thanas* of Sylhet district the floodplain area covers, parts of, three of these *thanas*: the entire Zakiganj *thana*, the part of Kanaighat *thana* south of the Surma river, and the part of Beani Bazar *thana* between the Surma and the Kushiyara. Of Kanaighat two unions are included: Paschim Dighirpar and Purba Dighirpar. Of Beani Bazar the entire Charkhai and Alinagar unions are included and about 30% of Dobhag and 60% of Sheola union. The total area of the selected floodplain area is approximately 400 km². The administrative units and the selected floodplain area are indicated in the map in Figure 4.3.

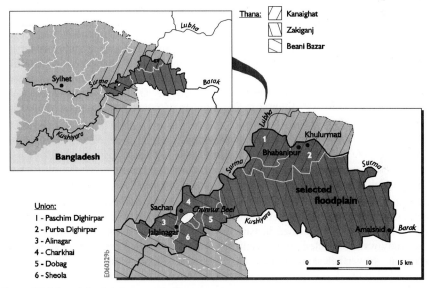

Figure 4.3 Selected floodplain area and villages

Within the selected floodplain area, four villages were selected. The reason to select entire villages instead of particular stakeholder groups, such as fishermen, was to ensure that less obvious stakeholders were included as well. Due to the morphological changes at the bifurcation point, the Surma is recently suffering from reduced low flows. For this reason, the focus in this case study is on villages along the Surma. Along the selected floodplain area, the Surma has one tributary: Lubha river. Two villages were selected upstream of this confluence, and two downstream. Because fishing and floodplain recession agriculture are reportedly the main income-generating activities related to the river ecosystem in this region, both up- and downstream of the confluence a village with many farmers and a village with many fishermen were selected. Village selection was done through a village survey, during which information was collected on the number of households, the main types of income-generation, and the level of development of the village. The level of development was assessed through information on type of houses, type of sanitary facilities and the availability of radios and televisions in the village. Average villages were chosen for the fieldwork. Upstream of the Lubha river, Khulurmati was selected as a village with many farmers and Bhabanipur as a fishermen village. Along the downstream section, the selected farming village is Sachan, while Jalalnagar was chosen as fishermen village. Table 4.3 lists the selected villages with the number of households in each village. The location of the villages is shown in Figure 4.3.

Table 4.3 Selected villages and number of households

Location/type	Selected village	Union - *Thana*	Nr of households
Upstream of Lubha			
• Farmers	Khulurmati	Purba Digirpar - Kanaighat	101
• Fishermen	Bhabanipur	Purba Digirpar – Kanaighat	46
Downstream of Lubha			
• Farmers	Sachan	Charkhai – Beani Bazar	79
• Fishermen	Jalalnagar	Charkhai – Beani Bazar	119

In villages in rural Bangladesh, houses of relatives are often built together to form a home (*bari*). Often, the material welfare of the households of the home is of a similar level, and many things are shared at this level. For this reason, at each home an interview was conducted with one of the households. Of the other households the information collected was limited to the number of household members, education level of the household head and income-generating and food-providing activities of the household. All homes of the villages were visited and interviewed, unless inhabitants of a *bari* were not willing to participate.

4.3.4 Limitations

The aim of this case study was to test and further develop the conceptual model and the approach to apply this model. The conceptual model and the stepwise approach to apply this model, as described in Chapter 3, were not available yet at the time this case study was conducted. As a result, the data collected do not entirely fit into that approach. Nevertheless, the experience gained in this case study has largely contributed to the development of the conceptual model and stepwise approach.

The focus of the case study (and in fact of this thesis) is on step 2, linking human well-being and the river ecosystem. Steps 3 and 4 are considered necessary to make the link with water resources management, but assessing the links in these steps is beyond the scope of this case study. For these steps, use needed to be made of analysis done by others. Since the research project 'Management of rivers for in-stream requirement and ecological protection' of the capacity building project focused on the relationships between river flows and river ecosystems, it was hoped that the results of that project could be used to fill in steps 3 and 4. However, the desired results of the research project were not available at the time of the analysis for this case study. Therefore, the links in these steps were estimated based on available data from reports and an existing hydrodynamic model application of Mike-11. It should be noted that the links of steps 3 and 4 are assumptions to show how to link the various steps of the conceptual model, but cannot be considered actual environmental flow requirements.

4.4 Identifying stakeholder groups

4.4.1 Numbers of stakeholders

Stakeholders are all people whose well-being is affected by changes in the flow regime of the Barak. The selected floodplain area is almost entirely flooded each year, except for the settlements on the higher elevated areas. Therefore, all inhabitants of this area can be considered stakeholders.

The total number of stakeholders is assessed based on the most recent population survey, of 1991. With this information, estimates are made of the size of the population and the number of households for the years 2000 and 2025 (Table 4.4). The growth rate of the population in Sylhet was 1.93% between 1981 and 1991(BBS, 1996). Growth rates for the entire population of Bangladesh are estimated at 1.32% from 2000 till 2025 (WARPO, 2000). Based on these numbers a growth rate of 1.7% was chosen to estimate the population of the floodplain area in 2000 (growth between 1991 and 2000) and of 1.32% to estimate the population in 2025 (growth between 2000 and 2025). Because the number of persons per household decreases, the number of households grows faster than the population itself. Between 1981 and 1991 the growth ratio for households was 2.05. Because other data on household numbers and size was not available, this growth rate is used to estimate household numbers in 2000 and 2025.

Table 4.4 Population and household numbers of selected floodplain area in 1991 with estimations for 2000 and 2025

	1991				2000	2025
	Zakiganj	Beani Bazar unions	Kanaighat unions	total floodplain area	total floodplain area	total floodplain area
population	174,038	51,239	27,222	252,000	294,000	408,000
households	29,836	7,980	4,578	42,400	50,900	84,500

4.4.2 Stakeholder groups

People are considered to belong to a different group, when changes in the flow regime are likely to affect them in a different way. Exactly how people will be affected became clear during the interviews. This section discusses a first division of stakeholder groups based on information on income-generation from the population census of 1991, and on location with respect to the river and related water bodies. In section 4.5 the need for identifying additional stakeholder groups will be discussed.

Main type of income-generation activity

BBS (1996) has listed the number of households per main activity. These activities can be grouped in agricultural activities, fisheries and other activities. The group 'other activities' consists of activities such as non-agricultural labour, handloom, business, construction, and transport. People in the floodplain area make use of various activities to generate income. In this thesis, households are indicated by their main activity. A farming household will receive most of its food and income through agricultural activities, but may practise other activities as well. This way, all stakeholders are divided in three groups: farmers, fishermen and others (Table 4.5).

Table 4.5 Percentage of households with different main income categories in the selected floodplain area in 1991 (BBS, 1996)

Type of income generation	1991			
	Zakiganj	Beani Bazar unions	Kanaighat unions	Total floodplain area
Farmers	58	31	67	54
Fishermen*	5	5	2	4
Others	37	64	31	41

* The census refers to this group as livestock/ forest fisher. There is no exact description. In this thesis the indicated number is considered a good estimate for the households involved in fishing as their main activity.

The 'other' activities are understood as not to depend on the river flow regime or on water in general. Agricultural activities and fisheries will require water, although this may be obtained from various sources of which the river is only one. Based on the census data it is estimated that the main income-generating activity is related to water for 58% of the population, and not related to water for 41% of the population. In the sections 4.5, 4.6 and 4.7 the relationships between these activities and the river will be discussed in more detail.

Location along the river

Because of confluences with tributaries, abstractions or intrusion of saline water, the location along the river may play a role in the changes in well-being experienced by inhabitants as a result of changes in the flow regime. Along the selected floodplain area, the Lubha river is the only tributary of the Surma. People living upstream of the Lubha river confluence may therefore be affected differently by changes in the Barak flow regime, than people living downstream of this confluence. Therefore the division based on income-generation can be further subdivided into upstream and downstream groups.

Location perpendicular to the river

People living further away from the river, may have a different relationship with the river flow regime than people living closer to the river. As was illustrated in Figure 4.2, in between the levees of the Surma and the Kushiyara, the area is bowl-shaped, with permanent water bodies in the middle. As a consequence, all villages are located on the levees, but some villages closer to the river itself, and others closer to the paddy fields and the water bodies on the floodplain. This location perpendicular to the river is not used to further divide the identified stakeholder groups. However, the selected village of Jalalnagar is located comparatively far from the river, and close to some *beels*.

4.4.3 Conclusion on stakeholder groups

The population of the selected floodplain area is now divided into six groups: upstream farmers, fishermen and other people, and downstream farmers, fishermen and other people. Table 4.6 gives an estimation of the number of households in each stakeholder group. For each group a representative village was selected, as indicated in the table. As Table 4.5 showed, the percentage of fishermen in Sylhet area is only 4%. The fishermen households often live together. As a result, there are villages with many fishermen and villages without any fishermen household. Certain villages consist of comparatively many farming households, but generally the villages consist of a mix of households with various types of income-generating activities. For this reason, both upstream villages and both downstream villages can be considered representative for the category 'others'.

Table 4.6 Stakeholder groups, percentage of floodplain population, total households in each group and representative villages

Group	% of population of selected floodplain area	Number 2000	Number 2025	Representative village
Upstream				
• Farmers	48	24,000	41,000	Khulurmati
• Fishermen	4	2,000	3,000	Bhabanipur
• Others	29	15,000	25,000	Khulurmati/Bhabanipur
Downstream				
• Farmers	6	3,000	5,000	Sachan
• Fishermen	1	1,000	1,000	Jalalnagar
• Others	12	6,000	10,000	Sachan/Jalalnagar

4.5 Relationship between human well-being and the river and floodplain ecosystem

4.5.1 Introduction

The previous section identified all people of the selected floodplain area as potential stakeholders in regard of changes in the flow regime. This section discusses for the three first order well-being values: income & food, health, and perception & experience, what the relationships with the river ecosystem are (the link) and what other factors influence this relationship (the context). The flow requirements to sustain the use of the river and floodplain ecosystem will be discussed in section 4.6.

4.5.2 Income & food
Link with river and floodplain

In chapter 3 the following links between the river ecosystem and income & food were mentioned:

- possibilities for income-generating activities;
- provision of food, comprising the three main nutrient groups of starch, proteins and vitamins; and
- provision of 'free' materials for various purposes.

Income-generating activities

To understand how income-generation is linked to the river flow regime, an overview is made of all income-generating activities carried out in the four selected villages.

Table 4.7 lists the number of households that depend on certain types of income generation. Various types of income generation are practised in the four villages, and many households have more than one type of income generation. This can have several reasons: 1) a person carries out several types of income generation, 2) one type of income generation is the alternative for another type of income generation in a certain season, or 3) there are more people in the household who contribute to the generation of income in different ways.

Income-generating activities related to water are indicated in Table 4.7 with an asterisk (*). From all the professions mentioned in Table 4.7, only six require water: agriculture, fisheries, day labour (since this mainly means agricultural labour), stone mining, laundry business and boating. Stone-mining, laundry business and boating were not mentioned in the classification of income-generating activities of BBS, and should therefore be considered as part of the category 'other activities'.

Of these, stone mining and laundry business do not depend on the flow regime of the Surma or the Kushiyara river. *Stone mining* depends on flash floods from the small steep rivers which come directly from the surrounding Meghalaya hills. The strong flash floods transport large cobble stones into Bangladesh. Various people work as a labourer in mining these stones, which are sold to be used in road and other construction works. These stones are not found in the less steep Surma and Kushiyara rivers and therefore, stone mining does not depend on the flow regime in these rivers. *Laundry business* was mentioned as the profession of only one man. The laundry is done in a pond near his house. Only in the dry season, when the pond water is insufficient or polluted, he makes use of the river water for his work. Although in this particular circumstance river water may be of importance, it was decided not to consider this activity as dependent on the Surma-Kushiyara flow regime. *Boat driving* is done during the wet season by men who have a profession as driver of a rickshaw or other public transport vehicle. According to the people engaged in this activity, driving over roads is preferred over driving a boat on the river. High river flows should therefore be considered a constraint to rickshaw driving, instead of a requirement for boat driving. Therefore, this activity is not taken into account as a Surma-Kushiyara dependent activity either.

Food

In rural Bangladesh, the diet of the population consists of three main items: rice, fish, and vegetables. While the cultivation of vegetables is mainly done on the highlands and rainfed, the production of rice and fish depends on the river flow regime. In the interviews, use of naturally occurring vegetation or of wild animals other than fish was not mentioned by the population.

Table 4.7 Number of households in the selected villages that depend on a certain type of income generation

Type of income generation	Bhabanipur 46 households	Khulurmati 101 households	Jalalnagar 119 households	Sachan 79 households
	No.	No.	No.	No.
Abroad	6	11	38	13
Agriculture *	9	36	39	26
Begging	2	1	4	1
Boat driving / transport *	2	2	3	1
Carpenter	1	0	0	3
Cook	0	1	0	0
Day labour *	5	24	2	11
Driver	0	0	2	2
Engine engineer	1	0	0	0
Fishing *	28	0	67	0
Imam	1	6	1	0
Laundry business *	0	0	0	1
Mason	0	0	1	2
Rickshaw puller	0	0	1	8
Selling land	0	1	0	0
Servant	2	2	2	0
Stone mining *	4	16	0	0
Tailor	0	3	0	0
Teacher madrasa **	0	4	0	0
Teacher school	0	4	2	2
Teacher mosque	0	1	0	0
Various business	1	11	4	27
Various paid jobs	0	2	0	8
Village homeopathic doctor	0	0	0	2

* activities are related to water
** *madrasa* is the Arabic word for school and is used in Bangladesh to indicate religious Islam-schools

'Free' materials

Natural vegetation can sometimes be used for various purposes such as construction of houses, manufacturing household utensils or as fuel. Use of such 'free' products contributes to the income & food value, because no income needs to be spent on purchasing these materials from other sources. This type of natural material use was hardly found in the selected villages. Various materials, especially for house construction or mat weaving, are bought from other areas. The only materials locally collected are various types of wood, leaves and cow dung which are used as fuel for cooking. However, according to the local population, these materials are collected from the higher elevated settlement areas, and not from the natural river floodplain. Therefore, these products are not considered to depend on the river flow regime.

Context

The following aspects of the context are relevant to understand the importance of river and floodplain related food and income:

- contribution of income from other sources;
- contribution of food or materials from other sources;
- poverty level;
- type of involvement in income-generating activity.

Each of these aspects is discussed in the following sections. The type of involvement in income-generating activity is discussed separately for involvement in agriculture and for involvement in fisheries/access to fish resources.

Contribution of income from other sources

Although people will often have a certain main income-generating activity, their total income may be based on a variety of activities. For each stakeholder group, Table 4.8 gives the average contribution of agriculture, fisheries and other activities to the total income. This table shows that although the main activity on average constitutes the largest part of the income (65-100%), other activities can still contribute up to 35%. This means that changes in the flow regime which affect agriculture or fisheries affect all people, but to various extents. This part of the context leads to a further specification of the relationship between certain ecosystem goods and services and the income & food component of the well-being of the stakeholders.

Table 4.8 Contribution of agriculture, fisheries and other activities to total income of stakeholder groups

Stakeholder group based on main income-generating activity		N	Average contribution of activities to total income of stakeholder group (%)		
			Agriculture	Fisheries	Other
Bhabanipur	Farmers	3	80	10	10
	Fishermen	27	4	94	2
	Others	16	11	0	89
Jalalnagar	Farmers	35	67	13	19
	Fishermen	52	2	94	4
	Others	32	5	3	92
Khulurmati	Farmers	34	91	0	9
	Fishermen	0	-	-	-
	Others	67	12	0	88
Sachan	Farmers	32	67	0	33
	Fishermen	0	-	-	-
	Others	47	2	0	98

Contribution of food or materials from other sources

No information was collected about import of paddy into or export out of the selected floodplain area. Therefore, the importance of local production of rice cannot be assessed. An alternative to fish from natural ecosystems may be fish from fish ponds. However, considering the large size of the population and the high poverty level, it can be assumed that the local food production plays an important role in the current situation. According to the National Water Management Plan (WARPO, 2000), fisheries provide 7% of the total protein intake, and is of crucial importance for the very poor as a source of nutrition and income. This part of the context does not lead to a further specification of the relationship with the river ecosystem, nor to a further division of stakeholder groups.

Poverty level

Different poverty levels may have a specific relationship with the type of income-generating activity people pursue as well as with their vulnerability to changes in resource availability. For the villages of Sachan and Jalalnagar a comparison was made of the poverty level and main income-generating strategy. A village meeting was organised to evaluate whether households were rich, middle class, or poor, according to their own perception. It should be noted that according to the people most villagers are poor, but that some people are comparatively better-off (classified as rich), while others are worse-off (classified as poor). The main characteristics attributed to the different classes by the villagers are given in Table 4.9, as well as the number of households in each category in the villages of Jalalnagar and Sachan.

Table 4.9 Characteristics and percentage of middle-class and comparatively rich and poor households by villagers of Jalalnagar and Sachan

Characteristic	Rich	Middle-class	Poor
Land tenure	Lot of land	Some land	No land
Employment	Business, highly paid job	Small business, normal job	Day labour, low-paid job, rickshaw driver
Extra income	Money from people working in European countries	Money from people working in countries in the Middle-East	-
Loan	-	Sometimes	All the time
Education	Secondary school*	Primary school*	No
Savings	A lot	A little	No
Housing	Some have house in a town		Some have no house
Miscellaneous			Sometimes they have no food Sometimes they steal
% Jalalnagar	15	30	55
% Sachan	10	44	46

* in Sachan people mention that the percentage of educated people is higher among middle class than among rich

The inventory in the village revealed that 51% of the population was classified as poor, 36% as middle-class, and 13% as rich. Table 4.10 shows that of the poor households, the largest percentage depends on fisheries. Middle-class households have other activities as their main income-generating activity, while the rich households mainly depend on agriculture. A high dependence of water for income-generation exists among both the poor and the rich households.

Table 4.10 Percentage of poor, middle-class and rich people with main income-generation activity

	N	Poor (51%)	Middle-class (36%)	Rich (13%)
Farmers	67	25%	38%	58%
Fishermen	52	46%	7%	4%
Others	79	30%	55%	38%
Total	198	100%	100%	100%

For the quantification of equity related to flow regime changes, it would have been useful to further divide the stakeholder groups based on poverty level. However, collecting this information in the various villages made assigning a poverty level to a household rather subjective. Therefore, this information was not collected for all households. Information on poverty level has not been used for further specification of the relationship with the ecosystem or for further division of stakeholder groups.

Type of involvement in agriculture

The previous section revealed that among the farmers, there are households in each of the three poverty levels. Farming is not an activity of only the rich or the poor; there are many different ways in which households can be involved in agriculture. Households can cultivate their own land, involve in share cropping, or work as a labourer or employee. The difference between labourer and employee is that an employee has a permanent position, while labourers are hired per day. Rich land owners can either hire employees or labourers or give their land for share-cropping. In a share-cropping system, a household cultivates land not owned by them and pays for this land use with a part (often 50%) of the yield. In Bangladesh, this share-cropping system is referred to as *barga*. In Bhabanipur and Khulurmati a little bit more than 30% of the households carries out agricultural activities. Approximately 25% is labourer or employee, another 25% cultivates under the *barga* system, while 35% of the households cultivates their own land. Of the remaining 15% of the households, 50% hires labourers or employees, while the other 50% gives land for *barga*.

As a result of the different types of activities, certain farmers are more vulnerable to losses than others. Day labourers and employees may not earn a high wage, but do not have to invest in seeds and chemicals and will depend less on the size of the yield. Households that cultivate their own land or involve in the *barga* system run the risk of losing their investments when the harvest is low or completely lost. Land owners are themselves often involved in other activities. They benefit from the agricultural harvest, but are not completely dependent upon it. These different levels of vulnerability associated with the type of activity are summarised in Table 4.11.

The various types of farmers have a different vulnerability to changes in the water resources system that affects agricultural revenues. Therefore, for the quantification of equity related to flow regime changes, it would have been useful to further divide the stakeholder groups based on type of involvement in agriculture. However, this information was not collected for all households. Therefore, information on type of involvement in agriculture has not been used for further specification of the relationship with the ecosystem or for further division of stakeholder groups.

Table 4.11 Income and risks associated with cultivation for different types of farmers and labourers

Type of farmer	Average size of land per person	Income	Vulnerability
Labourer	no land employed 1/ha	daily salary of 100 Tk* times number of days worked	middle (no investments, but poorest people and dependent on labour, agriculture largest employment sector for day labourers)
Barga	1-3 ha cultivation	½ yield of land minus investments	high (invest a lot, not many reserves)
Own land	3 ha	complete yield of land minus investments	high (invest a lot, may have some savings)
Having labourers	10 ha	complete yield minus investments minus wage of employees	middle (make investments, no need to hire many day labours when part of harvest damaged, rich so have reserves)
Giving for barga	10 ha	½ of yield of land	low (normally these people have their own business and do not invest in cultivation)

Type of involvement in fisheries/access to fish resources

Similar to the agricultural sector, there are different ways to be involved in fisheries. Most of the households that indicated to depend on fisheries for part of their income can be considered full-time professional, or traditional, fishermen. This is one of three types of fishermen distinguished by De Graaf et al. (2001), the other two being occasional fishermen; people who fish occasionally but relatively intensively during the period when fish is easily available, and subsistence fishermen; people who fish for their own consumption, often children and elderly people. In the selected floodplain area the professional fishermen constitute 4% of the population. Occasional and subsistence fishermen may comprise a larger group of people, which, however, fishes fewer days per year. In research on types of fishermen and total catch, in a different area of Bangladesh, De Graaf et al. (2001) found that between 1992 and 2000 on average 42% of the catch was caught by occasional fishermen, 25% by professional fishermen, and 33% by subsistence fishermen.

Occasional fishermen and subsistence fishermen are more difficult to identify, because they won't normally consider fishing part of their income-generating activities. Moreover, fishing used to be taboo for Muslims (De Graaf et al., 2001). Another reason why occasional fishermen cannot easily be found could be the fact that certain fishing activities practised by the occasional fishermen are illegal. The professional fishermen mentioned that various stretches of the river are fenced off by powerful people, which meant a restriction to their own fishing activities. In such a fenced-off river stretch, occasional fishermen set up a so-called pile fishery or *katha*. A *katha* consist of tree branches, often attached to sticks. The tree branches, which are located below the surface level of the river, provide shelter to fish. In a period of a few weeks to months, numerous fish find shelter between the branches. When after this period, a net is put around the *katha*, all gathered fish can easily be caught. SLI and NHC (1994a) mention that according to the Fish Act, pile fisheries are illegal in running waters to protect brood fish moving from the *haor* to the river to find over-wintering grounds and which may accidentally take shelter in the *katha*. However, they also mention that this is exactly what people do: they place the katha during flood recession in October/November and harvest in January. The fact that it was not possible to find any household in Khulurmati that admitted involvement in the *katha* next to their village, and which was harvested by Khulurmati people, can be considered evidence that occasional fishermen don't like to be associated with this type of fishing activity.

Another restriction to all fishermen is the system of leasing of the beels. Beels larger than 1.2 ha cannot be used freely but have to be leased. Leasing is done through auctioning. Beels larger than 8 ha are leased out by the Ministry of Land, beels between 1.2 and 8 ha at the *thana* level. Leasing can only be done by fishermen, but fishermen do normally not have enough money to pay the lease fee. Therefore, rich private money lenders lease the *beel* through a fishermen association. Leasing is normally done for a period of three years. Some leaseholders have the fish caught by the fishermen every year or every three years. Certain fishermen are employed as guards, for which they get either a monthly salary or are allowed to catch fish occasionally. In Jalalnagar most fishermen are a member of the Charkhai Fishermen Assocation which is the official leaseholder of Chunnur Beel. However, because the fishermen do not have sufficient capital themselves, a rich restaurant owner of Charkhai is the actual owner. This man invests in fingerlings, hires guards and organises fish catches. According to the professional fishermen, the person taking lease is the main beneficiary of the leasing system, and most of the professional fishermen hardly benefit from *beel* fisheries. According to the Fisheries Specialist Study of FAP 6 (SLI & NHC, 1994a) the leasing system encourages over-exploitation. Leaseholders who take a lease for one to three years consider it their right to catch all fish in the area. Fishermen get only a share of the profit, and to make this sufficient they try to make the total profit as large as possible. To protect the professional fishermen a policy was established in 1986 called New Fisheries Management Policy (NFMP). This policy is carried out by the Department of Fisheries in a number of *beels* (about 250 *beels* of the total 10,000 *beels*). There is not enough capacity in the department to maintain the policy in more *beels*.

During the monsoon season when the entire area is flooded, there are no distinct *beel* boundaries, and all people can use the area freely. This is most likely the period during which most of the subsistence fishing takes place.

Besides the actual fishermen there are some people who trade and sell fish. The people selling fish at the market have not caught the fish themselves but bought these from the fishermen or middle men. In the selected villages only few of such people could be found.

Similar to the households involved in agriculture, households in fisheries have different vulnerabilities with respect to changes in the flow regime. The professional fishermen are the most vulnerable because they are the poorest people and fisheries is their main, or only, source of income. When only fisheries in the river decline, the professional fishermen will be affected most severely. Degradation of *beel* fisheries, is more likely to affect the leaseholder. However, many fish species need both the *beel* and the river for their survival. The rich people, who take the lease for the *beel*, are again the least vulnerable, because the fisheries revenues constitute only part of their income. These different levels of vulnerability associated with the type of activity are summarised in Table 4.12.

Table 4.12 Income and risks associated with fisheries for different types of fishermen

Stakeholder	Fishing location	Fishing period/method	Income	Vulnerability
Professional fishermen	river *beel* when hired floodplain in monsoon season no fishing when house is flooded	all year/net	catch * price daily salary when hired as guard katha/lease system: 6/16 of total catch	high – poorest and powerless people depending on fisheries, but fishing grounds restricted
Subsistence fishermen	floodplain/river	receding floods	only own food	middle – catching own fish saves money
Occasional fishermen	floodplain	receding floods / katha	catch * price katha: 10/16 of total catch	middle – fishing is additional income
People taking lease (for 3 years)	*Beels*	once every three years – unsustainable methods like dewatering	fish yield – salary of guards- lease sum – investments in fingerlings	low – the people pay the sum for lease and invest in fingerlings. They normally are rich and have their own business.

Besides the professional fishermen, not many people mention fishing as an activity. Therefore, it was not possible to further divide the stakeholder groups based on the above description of type of involvement in fisheries.

Conclusions on income & food

The three main income-generating activities that can be distinguished with regard to use of the river ecosystem are agriculture, fisheries, and other, not water-related, activities. This section has discussed the importance of the various income-generating activities as well as the vulnerability of different groups of households within each main activity group. This contextual information is necessary to understand to what extent changes in the river flow regime can affect different groups of people. Because not all context information was available for all stakeholders, only the information on the contribution of income from other sources can be used to further specify the link between the income & food value and the river ecosystem. This information is summarised in Table 4.13 for the six identified stakeholder groups.

Although only few professions require river flows, 59% of the local population depends on these income-generating activities. Households with main activities not related to the river flow regime constitute a large part of the population as well: 41%.

Agriculture and fisheries are not only important for income for about 60% of the rural population, but also provide the main food items, and in this sense are probably important for all floodplain residents and probably for inhabitants of other regions of Bangladesh as well.

Table 4.13 Percentage of income obtained through various activities for each of the stakeholder groups

Stakeholder group	% of population	% of income obtained through		
		Agriculture	Fisheries	Other activities
Upstream				
• Farmers	48	90	1	9
• Fishermen	4	4	94	2
• Others	29	11	0	89
Downstream				
• Farmers	6	67	7	26
• Fishermen	1	2	94	4
• Others	12	3	1	96

4.5.3 Health

Link with ecosystem

In Chapter 3 the following links between the river ecosystem and human health were mentioned:

- availability of water of sufficient quality and quantity for drinking and sanitation;
- prevention of suitable habitats for disease-vectors;
- availability of medicinal natural vegetation;
- processing of waste;
- (prevention of) floods and droughts and related phenomena (e.g. sandstorms) (local climate and living conditions).

Availability of water of sufficient quality and quantity for drinking and sanitation

The water within the river provides a source of water for drinking and other domestic purposes for the people of the selected floodplain area. The importance of this source of water will be discussed in more detail in the context section. The availability of groundwater plays a role as well in providing water for domestic use. The possible link between flooding and groundwater recharge is not considered in this thesis. However, if river water levels and floodplain inundation are contributing to groundwater recharge, the main drinking water source, the ponds, also have some dependence on the river flow regime. According to the population, rainwater is sufficient to recharge the ponds and the river does not have a function in this.

Prevention of suitable habitats for disease vectors

In the current situation, vector-borne diseases were not mentioned as prevalent in the villages. However, when the flow regime changes, there is a possibility for disease vectors to develop. Quantifying this type of development requires a detailed assessment by health specialists, which was beyond the scope of this case study.

Availability of medicinal natural vegetation

Most of the villagers mention that when ill, they see a doctor and buy medicine from him. Very little natural vegetation is used for medicinal purposes. The few types of natural medicine used grow within the settlement area and are not considered to be part of the river ecosystem.

Processing of waste

Although some people mention that flooding takes away waste, many also tell that the floods bring dirt and debris from upstream areas. The role of the river in maintaining a clean environment is therefore undefined.

Prevention of floods and droughts

The northeast of Bangladesh is a wet area and more prone to floods than to droughts. The main life-threatening floods are the flash floods, which take place shortly after rainfall in the Meghalaya mountains. These floods are not related to the flow regime of the Surma and Kushiyara rivers, and cannot be enhanced or mitigated by upstream developments. Annual monsoon flooding is generally not life-threatening, although drowning does take place occasionally.

Concluding, the main link between the river flow regime and health in this region of Bangladesh is water for drinking and sanitation. In the next section the availability of alternative sources of water, and other ways to improve health will be discussed.

Context

According to the Bangladesh Bureau of Statistics, approximately 5% of the population of the floodplain area used the river as a source of drinking water in 1991. However, many more people may use the river as a source of water for domestic purposes other than drinking, if only during the dry season. In the villages of Sachan and Jalalnagar the river is not used much for domestic purposes. In the villages of Bhabanipur and Khulurmati however, use of the river water for domestic purposes is more common. A distinction can be made between people who use the river for all purposes and people who use the river only for some purposes, or only during certain seasons, and people who do not make use of river water for domestic purposes at all. Various households use pond water in the rainy season, when the river water is perceived to be polluted. In the dry season, at the other hand, the ponds have low water levels and the pond water is considered to be of poor quality, while the river water is perceived to be clear and of relatively good quality.

Various other sources of water are available in the area. BBS (1996) reports that 80% of the population makes use of ponds, 15% of tubewells, and less than 1% of wells or taps. Despite the presence of these other sources, for a small part of the population the river provides the only nearby accessible source. The selected upstream villages are located on the banks of the Surma river. It is expected that river use for domestic purposes is relatively high in these villages, compared to other villages on the floodplain. People who do not have their own water source, but live further away from the river, generally make use of common or other peoples' ponds and tubewells.

According to BBS (1996) the percentage of households making use of the river water is higher in the downstream parts of the selected floodplain than in the upstream parts

(Table 4.14). This is not in accordance with the findings in the selected villages. The selected villages in the downstream part may not be entirely representative of the floodplain area in regard of domestic water use. Jalalnagar is located too far away from the river for people to walk to collect water. Sachan is located close to the union centre, and is possibly a somewhat richer village than the other villages and has possibly more ponds and tubewells.

Table 4.14 Number and percentage of households in selected floodplain area using the river as source of drinking water

Part of selected floodplain	Total households (1991)	Households using river for drinking water (1991)	% of population
Upstream (Kanaighat unions and Zakiganj)	33,900	1,470	4
Downstream (Beani Bazar unions)	7,700	600	8
Total	41,600	2,070	5

The amount of water used for domestic purpose is low. Water is taken out of the river for purposes like drinking, cooking and cleaning. Other uses like washing and bathing are carried out in the river itself. The amount withdrawn is estimated at 10-20 l/p/d. For non-consumptive use the main requirement is that there is water flowing in the river. Naturally occurring deeper parts of the river can be used for bathing and washing.

Sufficient water of good quality is important to maintain human health. In the current situation, various diseases occur in the villages. Frequently mentioned were vomiting, diarrhoea and skin and eye infections. These diseases could be caused by water of poor quality or by poor personal hygiene, but may also have other causes, such as poor quality food.

Lower dry season river flows may increase the incidence of diseases related to drinking water quality and personal hygiene. However, since the diseases are already present in the current situation, maintaining the current flow regime will not help to improve the health situation. Other solutions are required for this. When further investigation would identify low use of water for personal hygiene as an important factor, improving access to the river for collecting water or for washing and bathing can be a solution. Nowadays in the dry season, the water level is very low compared to the levees and is difficult to reach. The construction of concrete steps and bathing sites can improve this situation. Another solution to improve personal hygiene could be a program to raise awareness on personal hygiene and sanitation and support the construction of sanitary type toilets (available to only 11% of the population in the selected floodplain area in 1991, versus 86% of other toilet types and 4% no toilet facility at all (BBS, 1996)). Alternative sources of drinking water close to home are a third type of solution. A complication in Bangladesh is the presence of arsenic in deep (20-100m) groundwater. Since the 1970s, tubewells were installed to reduce the incidence of water-borne diseases. However, in the 1990s arsenic poisoning resulting from drinking polluted groundwater became apparent (Pearce, 2001). Long-term consumption of arsenic polluted water can result in severe health problems, including cancer. Nowadays, it is estimated that 10 million tubewells pump water with high (up to 400 times the WHO standard) arsenic levels. Rainwater harvesting may form a suitable solution to this region with high precipitation. Water can be preserved in tanks for four to five months and used as a safe, easy to use source of water for mainly drinking and cooking. Initiatives to collect rainwater for domestic purposes are currently underway in Bangladesh (UNEP, 2002).

Conclusions on health

The main link between the river ecosystem and health is the provision of water for drinking and other domestic purpose. Analysis of the context has shown that the river water is of importance for drinking to only a limited number of households. Possibly, the number of households making use of the river water for some domestic purposes in specific seasons consists of a somewhat larger group of people. An estimation of the percentages of people making use of the river water for domestic purposes is given in Table 4.15. Although a decrease of river flows may worsen the health of a small part of the population, different types of measures than maintaining flows may be more effective in maintaining and improving human health, and can reach a larger group of people.

Table 4.15 Percentage of households depending on the river ecosystem for domestic water

Stakeholder group	% of population	Use of river water for drinking (%)	Use of river water for domestic purposes (%)
Upstream			
• Farmers	48	4	8
• Fishermen	4	4	8
• Others	29	4	8
Downstream			
• Farmers	6	8	16
• Fishermen	1	8	16
• Others	12	8	16

4.5.4 Perception & experience

Link with river

In Chapter 3 the following links between the river ecosystem and human health were mentioned:

- providing the scene for ceremonies;
- providing people with a sense of belonging to an area;
- providing opportunities for recreation; and
- preventing inconveniences, for example flooding of homesteads, difficult transportation.

Scene for ceremonies

With respect to experiencing religious ceremonies, stakeholders are all people who adhere to a religion which has rituals for which river water is used. Bangladesh is a Muslim country. In 1998, 83% of the population was Muslim, 16% Hindu, and 1% belonged to other religions (Buddhism, Christianity and Animists). The villages of

Jalalnagar, Bhabanipur and Khulurmati are completely Muslim. In Sachan, 19 of the 79 households are Hindu. The Hindu and Muslim people live in separate parts of the village.

The main ritual in the Islam religion which requires water, is the ritual bathing, which is required before each praying session. This ritual has no relationship with the river flow regime however, since any source of domestic water can be used for this.

Durgha-Puja, the main religious festival of the Hindu people has a relationship with the river. On the last day of this festival, which takes place in autumn, a statue of a goddess is immersed in the river. Other ceremonies involve watering of a special tree, the Tulsi tree, as well as of statues of gods or goddesses. This can be done by any type of domestic water, and has no relationship with the river ecosystem.

Sense of belonging to an area

In a situation where no changes have taken place yet, it is difficult to discuss with people the sense of belonging with regard to their surroundings. This aspect of perception & experience was therefore not considered in the case study.

Opportunities for recreation

In the villages in Bangladesh, the people are poor and do not have much time for recreation. The selected floodplain area does not offer many recreation opportunities for people from outside the area.

Inconveniences

Life-threatening floods were discussed in the section on health. However, while not dangerous, annual inundation is perceived as highly inconvenient by many villagers. Although land use is adapted to annually recurring floods, for many people floods are perceived as a nuisance. The village is flooded and roads cannot be used. People have a few options for travelling when the roads are flooded: they can swim or walk through the water or they use a boat. Travelling over road, either by foot or public transport, is by most of the villagers preferred over travelling by boat.

The houses of the poorer people in the villages are often located on the less elevated sites of the settlement. Inundation easily damages these houses built of mud. As a result, the inhabitants of these houses have to flee their houses during approximately one week per year or once every two years, and rebuild their house afterwards. To keep their children above the rising water levels in these houses without furniture, people use rafts within their houses.

Context

To protect the poor against the inconveniences of flooding, perhaps other measures than lowering the flood levels are possible. One can think of construction of dikes around settlement areas, or of supporting the poor to build elevated houses, and to help them to use other materials. However, it is not clear to what extent inundation originates from high river flows, or from local precipitation. When high precipitation is the main source, building dikes around the settlement will lead to drainage congestion and increase the problem.

Conclusions on perception & experience

Besides the Durgha-Puja festival, there are no ceremonies which require a certain river flow. The Durgha-Puja festival requires river flows of a certain, not assessed, discharge at the end of August, beginning of September. Approximately 16% of the population will benefit from this. The poorest part of the population suffers most from annual inundation of the villages. They have less access to boats, and their mud houses built in relatively low areas are highly prone to inundation. These results are summarised in Table 4.16.

Table 4.16 Percentage/ type of households for which perception & experience is linked to the river and floodplain ecosystem

Stakeholder group	% of population	Ceremonies	Flooding inconveniences
Upstream			
• Farmers	48	16%	All, but poor most affected
• Fishermen	4	16%	All, but poor most affected
• Others	29	16%	All, but poor most affected
Downstream			
• Farmers	6	16%	All, but poor most affected 16
• Fishermen	1	16%	All, but poor most affected 16
• Others	12	16%	All, but poor most affected 16

4.5.5 Conclusions on the relationship between human well-being and the river and floodplain ecosystem

After discussing the three first order well-being values, it can be concluded that the local population mainly uses river ecosystem goods and services to obtain income through agriculture and fisheries. For domestic water a small part of the population has no other source than the river, but only a small amount of water is needed for this. Natural vegetation and animals, except fish, are hardly used, or for some reason not mentioned. For personal transport roads are more important than water, and therefore high water levels cause problems. An overview of the findings for the three first order well-being values is presented in Table 4.17.

Table 4.17 Overview of stakeholder groups and importance of ecosystem goods and services

Stakeholder group	% of population	Income & food (% of income)			Health (% of population)	Perception & experience	
		A	F	O	Domestic use	Durgha Puja (% of population	Flooding inconvenience
Upstream							
• Farmers	48	90	1	9	8	16	All, poor most
• Fishermen	4	4	94	2	8	16	All, poor most
• Others	29	11	0	89	8	16	All, poor most
Downstream							
• Farmers	6	67	7	26	16	16	All, poor most
• Fishermen	1	2	94	4	16	16	All, poor most
• Others	12	3	1	96	16	16	All, poor most

Different stakeholders can be identified when looking at a certain type of use. For example in agriculture there are rich land owners and poor day labourers. The same holds for fisheries. However, this does not lead to different flow requirements, and to make a quick assessment of the required river discharge it is not necessary to consider all these different groups. On the other hand, if the flow required cannot be met, the different groups of stakeholder will experience a different severity of impacts. Although the collected data do not allow for a further subdivision of stakeholder groups, this difference in impacts will be mentioned were relevant.

4.6 Relationship between the river and floodplain ecosystem and the local flow regime

The main goods and services provided by the river ecosystem are the support of floodplain agriculture, the provision of fish and the availability of water for domestic purposes. This section discusses the link between the ecosystem goods and services and the flow regime for these three goods and services. In addition, the link between flow regime and the inconveniences caused by flooding of settlement areas is discussed.

4.6.1 Support of agriculture

Link with flow regime

Flooding of the floodplain area supports floodplain agriculture in two ways:

- sediment deposition; and
- water supply.

Sediment deposition

During the flooding of the floodplain areas, sediment is deposited on the fields. These sediment deposits lead to fertile lands. Although the function of flooding in fertilising the fields is recognised by the farmers, some farmers have the opinion that this leads only to marginally higher yields, which do not outweigh the negative effects flooding also has. The area has most likely not experienced long sequences of years without flooding, and no information is available about reduced fertility and yields, when the frequency of flooding decreases.

Because of the lack of data, the assumption is made that large floods are required once every 5 years to ensure that the fertile condition of the floodplain is maintained.

Water supply

The system of seasonal flooding has resulted in recession agriculture, where the receding floodwater fulfils the water requirement of the crops and perhaps induces other ecosystem processes which are beneficial for cultivation. The main crops cultivated in the selected area are the paddy varieties *aus*, *aman* and *boro*. Besides paddy, vegetables are cultivated within the settlement area and along the river banks. *Aus* paddy and the vegetables do not have a relationship with the flow regime: aus is a rainfed crop, and vegetables are irrigated from ponds and occasionally from the river. The water supply function of flooding of the floodplain area will mainly benefit the cultivation of aman and boro. The aman and boro varieties used in the villages require transplanting, and spend only part of the growing season on the floodplain.

Aman is first planted during the monsoon season on the high fields near the settlement. In September, when flood receding has started, the saplings are transplanted to the floodplain area. From these fields, the crop is harvested in November. With respect to flooding, the farmers of Sachan village estimated the need of approximately 15 cm water depth from land preparation in August till the end of September. During harvesting time there is preferably no water left on the fields. The total area used for aman cultivation on the selected floodplain area in the current situation is approximately 14,000 ha (BBS, 2001).

Boro paddy is transplanted onto the medium low land areas in December or January. During this period floods are still receding and these lower lying fields still have a layer of water standing on the field. An inundation depth of approximately 15 cm is required from land preparation stage at the beginning of December, until a few weeks before harvesting. Lower water levels in the initial stages require irrigation later on. Harvesting of boro rice takes place in April. The late harvesting time of boro is a problem in areas prone to flash floods. The total area used for boro cultivation on the selected floodplain area in the current situation is approximately 12,000 ha (BBS, 2001).

Both aman and boro paddy can be damaged when the spikes touch the water, which means that too high water levels need to be avoided. When inundation lasts too long, the softened soil may not provide sufficient support for the plant roots.

Context

The following aspects of the context can be identified:

- alternatives for sediment deposits;
- importance of current cropping scheme;
- possibility of artificial regulation of water levels;
- importance of flooding from the river versus inundation through local precipitation.

Alternatives for sediment deposits
Possibly, the use of artificial fertilisers may be able to replace the fertilising effect of the sediment deposits. In this research, the potential of using fertiliser has not been investigated. The costs and benefits, and the risks of side-effects through pollution should be further investigated.

Importance of current cropping scheme
The cultivation of aman and boro paddy on the selected floodplain area poses certain requirements with respect to the flow regime. However, the reason that aman and boro are cultivated is because these are the crops that match best with the natural circumstances. This could imply that in case the circumstances change, the area may no longer be suitable for aman and boro cultivation, but may have become suitable for the cultivation of other crops. Such shifts to other crops can only be realistic when structural changes in the flow regime can be expected. Farmers may be reluctant to change to something they are not used to, which means that a shift like this should be well-guided. From this perspective, changes in the flow regime are not necessarily negative.

Possibility of artificially regulating water levels
The opposite of adjusting to a different hydrological situation is to control the water levels completely by constructing full flood protection with drainage and irrigation facilities. Such controlled systems can increase crop yields and provide certainties, but will be expensive to construct, manage and maintain. Without natural flooding, fertiliser will have to be applied, which leads to increased cultivation costs. Moreover, the construction of irrigation facilities requires interventions which may have negative impacts on the natural system.

Importance of flooding from the river versus inundation through local precipitation
The requirements for water levels on the field cannot directly be translated into a flow regime in the river. Local precipitation is high, and according to some of the villagers, flooding is not required to have the crop water requirements fulfilled. This topic requires further investigation.

Conclusions on local flow requirements to support agriculture
In the current situation, annual inundation supports aman and boro cultivation. To maintain the fertilising function, flooding is required regularly, but not necessarily every year. These requirements are summarised in Table 4.18. Analysis of the context shows various alternatives for the current cultivation system, as well as uncertainties in assessing the importance of the current cropping scheme and of the role of local precipitation.

Table 4.18 Local flow requirements to support floodplain agriculture

	Inter-annual	Intra-annual											
		J	F	M	A	M	J	J	A	S	O	N	D
Aman Cult.	Freq. of large floods: 1/5 years	15 cm, or water in beel for irrigation				Seasonal inundation required to saturate soils				15 cm			
Boro Cult.													15 cm

4.6.2 Fish
Link with river flow regime
River flows are important to sustain fisheries in both rivers and *beels*. An often heard distinction of fish types in Bangladesh is between large fish (*boro mach*) and small fish (*choto mach*). Most large fish species are known to spawn in rivers, whereas small fish spawn in the floodplains. A distinction which is perhaps more scientific, but possibly comparable, is in white fish and black fish (De Graaf *et al.*, 2001):

- *"white fish:* these fishes migrate upstream and laterally to the inundated oxbow lakes and floodplains adjacent to the river channel in the late dry season or early rainy season in order to spawn in the quiet sheltered and nutrient rich waters. The eggs and newborn larvae of these species are transported passively by the flood into the floodplain areas, where the larvae feed on developed plankton. At the end of the rainy season, the adults and young of the year escape to the main channel and most likely the deeper *beels*, in order to avoid the harsh conditions of the floodplain during the dry season. This group is also referred to as riverine fish.
- *black fish:* these fishes are mainly omnivorous/carnivorous bottom dwellers. They reproduce at the onset of the pre-monsoon when the water level in the *beels* starts rising due to the congestions of rainwater. At the end of the rainy season the young of the year and adults migrate back to, or get trapped in, low-lying beels, where they can survive the harsh conditions of these permanent water bodies during the dry season. They are adapted to resist low dissolved oxygen concentrations and high water temperatures. The main adaptation is their auxiliary respiratory organ used for the uptake of atmospheric oxygen. This group is also referred to as *beel* fish."

To estimate the flow requirements for fish, use is made of the four distinct seasons of the fish life cycle, identified in the Fisheries Specialist study of FAP 6 (SLI & NHC, 1994a):

1. overwintering dry season: December to March;
2. spawning migration season: April to June;
3. nursery/grazing season: June to September;
4. flood recession season/returning: September to December.

Overwintering takes place in river *duars* (deeper permanent water bodies in rivers) and *beels*. Many fish species are not very selective about where to spend their winter. Sufficient water is required to maintain temperatures and oxygen concentration respectively below and above critical levels. Near the selected floodplain area, 5 duars with depths ranging from 12-30 m where found in the Lubha river, and 7 duars with depths ranging from 6-9 m in the stretch of the Surma between Amalshid and Sylhet (SLI & NHC, 1994b). However, because of longitudinal migration, fisheries in the selected floodplain areas may depend on *duars* elsewehere.

Many *beels* dry up each year naturally, others are drained for agricultural purposes, or dewatered to catch all fish. According to SLI and NHC (1994b), Sylhet district does not contain important fish producing *haors*. However, the district does count 1,084 *beels* with a total area of 118.7 km^2, of which approximately 58% is perennial. This means that in the entire Sylhet district, the perennial *beel* area is around 6900 ha. Because no detailed information is available for the selected floodplain area, the ratio between perennial *beel* area and total area of Sylhet district is applied to estimate the perennial *beel* area in the selected floodplain area. This leads to a required perennial *beel* area for the selected floodplain of 850 ha.

Because fish species have a preference for *spawning* either in the river or on the floodplain, many fishes need to migrate from their over-wintering ground to their spawning place. This requires sufficient water levels to connect the different water bodies and sufficient inundation of the floodplain to allow spawning, as well as a timely opening of the *khals*. Back migration to the over-wintering grounds will presumably be possible when the required flooding for spawning and grazing has taken place.

The fingerlings hatched from river breeding species need to get up to the floodplains for *grazing*. They are brought there passively with flood water or through *khals*. All fish species use the floodplain for grazing, which means that a certain magnitude of inundated floodplain area is required to provide food for all fish. An estimation of the required flooding extent can be made with a formula which links catch and flooded area (SLI & NHC, 1994a):

$$C = 4.23 \, A^{1.005}$$
C = catch in tons/year
A = area in square kilometres

For the total northeast region, the nominal capture fisheries production is between 80,000 and 90,000 tons/year, which means a yield per ha/year of 37-45 kg. The northeast region has 21.710 square km of flood prone land. With this formula this would mean a maximum total yield of 96,500 ton. To calculate the flooding extent required to feed a certain population, use is made of the assumptions by SLI and NHC (1994a) that the per person consumption is 6 kg per year, and that an average household consists of 5 members. Moreover they assume that 85% of the caught fish is consumed in the region.

With these assumptions, the required fish catch for the selected floodplain area, with a total population of 294,000 in 2000 is 2,100 ton/year. The floodplain area required for this would be 48,100 ha. This area is larger than the total area of the selected floodplain. Possibly, part of the fish consumed in the area is imported from other regions. The western part of the northeast region has deeper floodplains and a larger area of perennial water bodies, than the eastern part where the selected floodplain area is located. A large part of the total fish catch may take place in the haor area of Sunamganj, the district west of Sylhet, with flooding depths up to 6 meters and more. Therefore, we assume that only 90-95% of the consumed fish catch is caught locally, which requires a floodplain inundation of approximately 38,000 ha.

In the *flood recession season* fish return from the floodplain to the deeper water of *beels* and rivers. For the movement to the river open *khals* containing water (or structures enabling fish passage) are required.

The fishermen themselves have other ideas about the flows needed for fishing: too low water levels make boating difficult, while during high water levels it is difficult to catch fish, because their nets cannot reach the river bed. Moreover, fishermen are often part of the poorest people in the community who live on the less elevated parts of the settlement. During large floods the fishermen are occupied with surviving and are not able to go fishing. Therefore, they prefer constant average water levels. Nevertheless, the fishermen are aware of the importance of floodplain inundation and of a connection between rivers and *beels*, since this is the main point of dispute between boro farmers who want to keep their fields dry until the crops have been harvested, and the fishermen who want to open the connecting *khals*.

Context

River flow is not the only factor important to sustain the fish population: over-fishing and unsustainable fishing methods such as de-watering, reduced water quality, and fish diseases are some of the factors contributing to a reduced fish population (SLI & NHC, 1994a). Another threat to the fish population is the disappearance of the *duars*. These deeper permanent water bodies in the rivers are disappearing as a result of sedimentation, which poses a threat to the fish population. This phenomenon is however mainly mentioned for the small fast flowing tributaries, and do not have a relationship with the flow regime of the Surma and the Kushiyara rivers.

Cultivation of fishes in ponds may seem a suitable alternative when river flows no longer allow for fisheries. However, to construct a pond people need to have land, which is not the case for most of the professional fishermen. Hence, fish cultivation in ponds is mainly carried out by the comparatively rich.

Conclusions on local flow requirements to provide fish

Based on the discussion in this section, flow requirements can be estimated for the four seasons which are important in the fish life cycle. These requirements are summarised in Table 4.19.

Table 4.19 Local flow requirements for the provision of fish

	Inter-annual	Intra-annual									
		J	F	M	A	M	J	J	A	S	O	N	D
Fish	-	overwintering: sufficient water in beels and duars			spawning: links between river and beels			grazing: large floodplain inundation		return: connection to rivers and beels			over wintering

4.6.3 Availability of water for domestic use

Link with the river flow regime

As was discussed in section 4.5.3, the quantity of water required for domestic use is low. Since no storage of water is possible, a constant supply is required to ensure daily use. All of the people who make use of river water for domestic purposes have the same opinion about the quality of the water: the quality is poor in the rainy season, while in the dry season, the water is clear although the flow is low. During high discharges the river has high sediment levels, and according to the population the river also brings waste, including dead cows and goats. Despite this perceived pollution, most of the people who make use of the water for drinking, drink the water directly, without any treatment. No information on levels of bacterial pollution in the different seasons is available. Therefore, it remains unclear whether pollution levels are only perceived, or whether actual pollution exists.

Context

Instead of adjusting flow regime levels to maintain water quality, the river water could be treated. Some households mention that they add a chemical substance called *fitkuri* to the pond water before drinking. Other aspects of the context, such as improving access or providing alternative sources for domestic use, were already discussed in section 4.5.3, and are not repeated here.

Conclusions on local flow requirements for the availability of water for domestic use

The only flow requirement to maintain the current use of the river water for domestic purposes is that the flow should not become zero. The water should remain flowing.

4.6.4 Local flow requirements and flooding inconveniences

For peoples' perception of comfortable living circumstances it is desired that the settlement area is not flooded. This means that the current high flows need to be restricted. The elevation varies within the settlement area, and houses are generally constructed on the more elevated parts. Assumed is that inundation depths of the settlement area should not exceed 30 cm. The total settlement area in the current situation is approximately 17,000 ha (BBS, 2001). The context for this issue was discussed already in section 4.5.4.

4.6.5 Conclusions on the relationship between the river and floodplain ecosystem and the local flow regime

This section discussed the type of flow regime requirements related to the use of the river ecosystem. As was discussed in section 4.3, the link between the local flow regime and river ecosystem is a necessary step in linking water resources management and the human well-being benefits of river ecosystems, but the actual assessment of this link lies outside the scope of this study. Data and expertise in the field of ecology are required to assess this link more accurately. Therefore, the requirements summarised in Table 4.20 are meant to provide a general impression of during what season, or how frequent a certain type of river flow may be required.

Table 4.20 Monthly flow requirements for different well-being components

Well-being	Eco system G & S	Inter-annual	Intra-annual										
			J	F	M	A	M	J	J	A	S	O	N	D
I&F	Aman Cult.	Freq. of large floods: 1/ 5y					Seasonal inundation required to saturate soils				15 cm		
	Boro Cult.		15 cm, or water in beel for irrigation											15 cm
	Fish		overwintering: sufficient water in beels and duars			spawning: links between river and beels			grazing: large floodplain inundation			return: connection to rivers and beels		overwintering
Health	Dom. use		A constant flow of water is required, stagnant water should be avoided. Discharge can be low.											
P&E	Flooding		Inundation levels of settlement areas should not exceed 30 cm all year.											

4.7 Relationship between the local flow regime and the upstream flow regime

4.7.1 Link with upstream flow regime

The flow requirements at the selected floodplain area need to be translated into a flow requirement at an upstream location where discharge regulation takes place. Normally, this location would be the reservoir from where releases are to be made, in this case the Tipaimukh dam in India. However, in transboundary rivers, the information required is rather the flow regime that should cross the border. Therefore, this section discusses the flow regime upstream at Amalshid (see Figure 4.3), where the Barak river enters Bangladesh and bifurcates into the Surma and the Kushiyara rivers. Besides for the dialogue with India, Bengal authorities can use this information to decide on the need for construction works to stop or reverse the natural morphological changes at the Surma-Kushiyara bifurcation point within Bangladesh.

The flow parameters of interest for the selected floodplain area were discussed in the previous section (Table 4.20) and are the discharge in the Surma (important for fisheries and domestic use), the inundation depth and extent of the floodplains (important for paddy cultivation and fisheries), and the flow between rivers and floodplain. These three parameters need to be linked to the upstream flow regime. For this link use is made of an existing 1-D Mike-11 model for part of the northeast region of Bangladesh, with simulation results available for the period from April 1, 2000 till March 31, 2001. Plotting Barak discharge for 2000-2001 against longer series based on 30 years of measurement (SLI & NHC, 1993c) (see Figure 4.4) showed that 2000-2001 can be considered a relatively wet year.

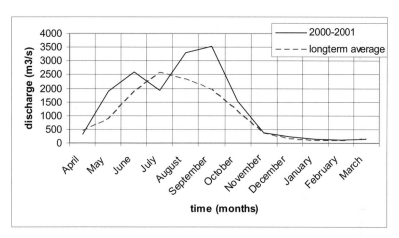

Figure 4.4 Hydrographs of 2000-2001 Mike-11 Barak inflow and of 30 years measured Barak inflow

For the river sections and the connecting *khals*, discharges are directly available from the existing results file. However, during low Surma flows, water that enters the Surma from the Lubha may flow in the opposite direction towards the bifurcation point. Because this leads to negative results for the Surma discharge, the absolute value of the simulation results is used. To calculate inundation depth and extent for the floodplains, some post-processing is done. In this model, not only the rivers, but also the floodplain areas are represented through 1-D branches. The floodplains are connected to the rivers through weirs, over which water flows when a certain discharge threshold is exceeded. Rainfall provides an additional inflow of water onto the floodplains. At the downstream end of the floodplain, a branch connects the floodplain with the Kushiyara river, through which drainage can take place.

From the water level simulation results for the floodplain branch, the inundation depth and extent are calculated as follows. The model contains in the selected floodplain section six cross sections at distances of 0, 8500, 20800, 34100, 39100 and 48000 meter from the start of the branch. These cross sections contain a relationship between level and cumulated storage width of the branch at that location. This means that the water level results can be transformed into width of the water surface. These cross sections are assumed to be representative for a length from halfway between the location of the cross section and the location of the previous cross section till halfway the distance to the next cross section. For the first and last cross section, the location of the cross section itself was considered respectively the start and end of the representative length. The surface area of the floodplain could then be calculated through multiplying length and width for each of the sections. The maximum surface area, at maximum elevation of the cross section, was with this method estimated at 40,400 ha. The total area of the floodplain according to the Bangladesh Bureau of Statistics is 43,000 ha. The deviation of 6% between estimation and statistics is considered acceptable.

The information on elevation and surface area was used subsequently to obtain the number of ha with a certain depth, averaged to monthly values:

- From the average monthly water level of a floodplain stretch, the levels corresponding with inundation depth of 10 cm, 20 cm, etcetera are calculated.
- For each depth class the total storage width is calculated with the elevation-storage width relationship belonging to the cross section of the floodplain stretch.
- The area in a depth class is calculated by subtracting for the first class the calculated storage width from the maximum storage width (this area is dry), and multiplied with the length of the floodplain stretch. Subsequently, the number of ha in the next depth class is calculated by subtracting the storage width of that level from the storage width of the previous level, and multiplying with the length. This way all areas with a certain inundation depth are calculated.
- Summation for all floodplain stretches gives the total area in a depth class per month, for the particular average monthly water level.

The result of these steps for each level of floodplain inundation is the area inundated with a certain depth.

4.7.2 Context

The context for the link between upstream and local flow regime refers to other factors than hydraulic laws which affect this relationship. In the Surma-Kushiyara floodplain, farmers regulate the inflow through and drainage from the floodplain area through artificially opening and closing of the *khals*. This means that inundation may last longer than simulated with the model, and high river flows alone cannot guarantee a connection between river and floodplain. Because of the ad-hoc character of these regulations, this cannot be taken into account in the available 1-D model application.

4.7.3 Conclusions on the relationship between local and upstream flow regime

This section discussed how upstream flows can be linked to the local flow regime, and what other factors need to be taken into account. With this relationship the required local hydraulic parameters can be assessed for any simulation results from the Mike-11 model or a similar model. Because of the limited data availability, in this case study this was done for the year 2000-2001 only.

4.8 Combining all: impacts of flow regime changes on human well-being

The previous section discussed parts of the relationship between the river flow regime and human well-being. This section combines all steps, and discusses what impacts the various stakeholder groups may experience when the flow regime changes.

4.8.1 Human well-being and flow requirements

With the links between human well-being, the river ecosystem and the river flow regime, this section estimates the extent to which water and flow requirements are met in the current (approximately the year 2000) situation. For this purpose, it is assumed that the goods and services of the current situation are sufficient for the current stakeholders, and that flow requirements for the stakeholders should fulfil the requirements of these goods and services. To evaluate whether requirements are met, the flow requirements of current floodplain use for agriculture, fisheries and settlement, as well as current river flow requirements for fisheries and domestic use are compared with the processed simulation results for the river and the floodplain in 2000-2001. The flow requirements for the relevant ecosystem goods and services are either related to floodplain inundation, to river depth and discharges or to discharge through the connecting *khals*. Inter-annual requirements for large floods cannot be taken into account, because the results for the local flow regime are available for only one hydrological year. This section discusses first the requirements with respect to the floodplain, and subsequently with respect to the river flow requirements.

Floodplain requirements

According to the Census for Agriculture (BBS, 2001), 14,000 ha of the selected floodplain is used for aman cultivation and 12,000 ha for boro cultivation. 3,000 Ha is used for aus and vegetable cultivation. Corrected for the cropping intensity of 1.3, 22,000 ha is net cultivated in the selected floodplain area. This is a little lower than the 67% net cultivated area that SLI and NHC (1994c) give. Of the 18,000 ha left for other uses, 850 ha was assumed to be perennial water bodies, leaving 17,000 ha for settlement and natural vegetation. Each type of floodplain use requires a certain area with a certain inundation depth in a certain month. The requirements for inundation depths are displayed in Table 4.21 in terms of inundation depth classes. Cultivation of aus and vegetables requires low inundation depths similar to the requirements of the settlement area. For this reason the area for aus and vegetables are combined with the settlement area, leading to a total area of 20,000 ha with a requirement of maximum 30 cm inundation all year round.

In the estimation of the extent to which flow requirements are met, it is assumed that all land use types take place in distinct areas, except for the large floodplain inundation for fish. This means that aman land can only be inundated, when the perennial *beel* area and the boro area are inundated as well. Therefore, the required inundated area for aman cultivation is increased with the boro and perennial *beel* area. Similarly, the required inundated area for boro cultivation is not only the boro area, but also the perennial *beel* area. In the estimation of meeting flooding requirements, the error made by not taking into account double-cropping in certain areas is considered small, and therefore ignored.

Table 4.21 Inundation requirements for the identified floodplain uses

Month \ ha	Aman	Boro	Aus, vegetables and settlement	Fish	
	14,000+12,850 =26,850	12,000 + 850 =12,850	20,000	850	38,000
	Inundation requirement (cm)	Inundation requirement (cm)	Inundation requirement (cm)	Inundation requirement (cm)	Inundation requirement (cm)
April		0-10	0-30	> 10	
May			0-30	> 10	
June	> 10	> 10	0-30		> 10
July	> 10	> 10	0-30		> 10
August	> 10	> 10	0-30		> 10
September	> 10		0-30	> 10	
October	0-10		0-30	> 10	
November	0-10		0-30	> 10	
December		10-30	0-30	> 10	
January		10-30	0-30	> 10	
February		10-30	0-30	> 10	
March		0-10	0-30	> 10	

With the estimation of area per depth class and the requirements of Table 4.21 the amount of land suitable for a certain type of use can be derived. Comparing this with the current land use in the floodplain shows whether the simulated discharge is sufficient. The simulation results and comparison with the requirements are both displayed in Table 4.22.

Table 4.22 Simulation results (2000-2001) and degree to which inundation requirements are met

Month	Area with required inundation depth (ha)				% to which requirements are met			
	0-10 cm	0-30 cm	10-30 cm	10 cm-max	Aman	Boro	Aus, vegetables and settlement	Fish
April	33,000	36,600	3,600	7,400	N/a*	>100	>100	>100
May	27,100	29,500	2,400	13,300	N/a*	N/a*	>100	>100
June	26,400	28,400	1,900	14,000	50	>100	>100	35
July	26,400	28,400	1,900	14,000	50	>100	>100	35
August	22,600	24,700	2,100	17,800	65	>100	>100	45
September	21,000	23,200	2,100	19,400	15	N/a*	>100	>100
October	28,100	31,400	3,300	12,300	>100	N/a*	>100	>100
November	33,800	37,300	3,500	6,600	>100	N/a*	>100	>100
December	35,500	38,600	3,000	4,900	N/a*	25	>100	>100
January	36,200	39,000	2,800	4,200	N/a*	25	>100	>100
February	36,400	38,900	2,500	4,000	N/a*	20	>100	>100
March	35,100	38,500	3,400	5,300	N/a*	>100	>100	>100

* N/a: not applicable; there are no inundation requirements for this month

Table 4.22 shows that in September only 15% of the aman area has the required inundation depth. Also, during the transplanting and growing period of boro, water has receded far, leaving only 20-25% of the boro area with the required inundation depth. Agricultural yields may be at a reduced level during the year 2000-2001. At the other hand, the settlement area did not suffer from flooding in the year 2000-2001. The floodplain area inundated for fisheries is only 35-45% of the area required, which may result in reduced fish yields that year. These results are not in concordance with the identification of the year 2000/2001 as a relatively wet year. However, as was mentioned in section 4.3, the links between river flows and goods and services of the river ecosystem are not the main focus of this case study and based on many assumptions. The results should be regarded as illustration of the applied approach.

River flow requirements

Both fisheries and domestic use of river water pose requirements to the local flow regime. Fisheries require deep *duars* during winter and timely overflow from the river through the *khals* into the floodplain. Domestic use requires flowing water all along the length of the selected river stretch. The requirements for the river are summarised in Table 4.23. Table 4.24 shows the simulation results for the selected river flow parameters (depth, overflow discharge and river discharge), as well as the extent to which the requirements are met.

Table 4.23 Depth and discharge requirements for identified river uses

Month	Requirements		
	Duars (7 areas) (m depth)	Khals (m^3/s)	Domestic use (m^3/s)
April			>0
May		>0	>0
June		>0	>0
July			>0
August			>0
September			>0
October			>0
November	>6		>0
December	>6		>0
January	>6		>0
February	>6		>0
March	>6		>0

According to Table 4.24 the requirements for the duars are met for approximately 70-80%. A limitation in the simulation results is that the depths given are the averages over a river stretch. Deeper spots along such a stretch can therefore not be identified from the simulation results. The requirements for the connection between the river and the floodplain are met, as well as the requirement for domestic water use of the river water.

Table 4.24 Simulation results (2000-2001) and degree to which requirements are met

Month	Results 1-D simulations			% to which requirements are met		
	River depth (m)[1]	Overflow discharge (m^3/s)[2]	River discharge (m^3/s)[3]	Duars	Khals	Domestic use
April	6	1	53	N/a*	N/a	>100
May	10	141	617	N/a	>100	>100
June	12	192	890	N/a	>100	>100
July	11	69	662	N/a	N/a	>100
August	13	408	1,182	N/a	N/a	>100
September	13	472	13,151	N/a	N/a	>100
October	9	59	5,224	N/a	N/a	>100
November	5	0	51	85	N/a	>100
December	4	0	3	65	N/a	>100
January	4	0	2	65	N/a	>100
February	4	0	2	65	N/a	>100
March	4	0	2	65	N/a	>100

1. depth is the maximum of average monthly river depth along the selected river stretch of the Surma
2. overflow discharge is the sum of average monthly overflow at the four weirs which connect rivers and floodplain
3. river discharge is the minimum of average absolute monthly discharges along the selected stretch of the river.
* N/a: not applicable, there are no requirements for this month

Discussion of the current situation

For the current situation with respect to cultivation the focus was on total land available with the inundation pattern required for a particular crop. However, in practise, these requirements are more location specific. When the inundation pattern on an aman field is suitable for boro, an aman farmer may not easily shift to cultivate boro instead of aman. When changes in inundation patterns are more structural, however, structural shifts to other crops may occur. Also, farmers are to a certain extent able to regulate the inundation depth on their fields, through opening or closing of the *khals*. This may mean that the main requirement is sufficient flooding from June till August, and that the rest is up to the farmers themselves.

The regulation of water levels by farmers may be a cause of conflict, since fishermen require timely opening of the *khals*, to enable the migration of fishes. While the simulation results show a discharge from rivers to the floodplain in May and June, in reality this depends on local regulation.

Although not all requirements are met in the year 2000-2001, it is assumed that if a longer timeseries of river discharge data would be considered, the requirements in the current situation are met, since the current use of ecosystem goods and services is the result of the current flow regime. The important question to ask is: what will be the impact on the well-being of the stakeholders of changes in the flow regime? Although, for example, all farmers have the same flow requirement with respect to inundation depths for aman and boro, the large differences between farming households may lead to different degrees of impact when the flow requirements for aman and boro are no longer met, as was discussed as part of the context in section 4.5. Possible impacts of flow regime changes for the identified stakeholder groups are discussed in the following section.

4.8.2 Impacts of changes in the flow regime

This section discusses the potential impact in terms of direction, order of magnitude and stakeholder groups affected, for the following three types of changes:

- operation of the Tipaimukh reservoir for hydropower;
- diversion of water for irrigation of the Cachar Plain;
- morphological changes at the Surma-Kushiyara bifurcation point at Amalshid.

The *Tipaimukh* reservoir is planned to be a multipurpose reservoir for hydropower-generation, flood control, and irrigation. This paragraph discusses the impacts of hydropower-generation and flood control; irrigation is considered separately in the next paragraph on the irrigation of the Cachar plain. Generally, hydropower dams flatten the intra-annual river flow regime: peaks will be reduced, while low flows will be increased. Flood control has similar impacts: high flows will be stored in the reservoir and released during the dry season. SLI and NHC (1993a) estimate that peak flows at Amalshid will be reduced with 25%, and flood water volumes with 20%. The dry season volume is expected to increase with 60%. For floodplain cultivation, this means that the large floods which fertilise the floodplains may no longer take place. Also, the inundation depth, extent and duration may decrease. However, because of high local precipitation, there is probably still a large amount of water available to cultivate part of the area. Different inundation patterns may require a shift to different crops. Fisheries may be affected when the connection between the rivers and the floodplain may no longer take place, or at other times. Also, a smaller inundated area may reduce the amount of food available for the grazing fishes. For domestic use of river water the situation improves (provided that the released water is of good quality): domestic requirements are constant over the year and will benefit from regulated flows. Because high flows are reduced, less flooding of the settlement area may occur, improving the perception & experience part of the well-being of the local communities. Especially the poor will benefit from this.

Irrigation of the Cachar Plain will be a further use of the Tipaimukh reservoir. In this case, the flow regime is not only regulated, but the actual amount of flow is reduced. It depends on the type of irrigation, whether the decrease takes place during the dry or the wet season. In the wet season, irrigation may be required as supplementary irrigation only, requiring only a small portion of the high wet season discharge. When cultivation is planned for the dry season however, a much larger portion of an already lower discharge may be used. In addition to the impacts of use of the Tipaimukh reservoir for hydropower, this may mean that the negative impacts on agriculture and fisheries are further worsened, and that the positive impacts of regulated low flows for domestic water are diminished. Because high flows are reduced, less flooding of the settlement area may occur, improving the perception & experience part of the well-being of the local communities. Especially the poor will benefit from this.

The current trend in the *morphological changes at the Surma-Kushiyara bifurcation point* leads to a reduction of the Surma inflow and an increase of Kushiyara inflow. This reduces the availability of river water for domestic purposes along the Surma during the dry season, especially in the upstream reaches before the confluence with the Lubha river. For floodplain cultivation and the fish life cycle stage at the floodplain, it does not matter whether flooding takes place from the Surma or from the Kushiyara river. The fishermen along the Surma, however, will be negatively affected when water levels become too low in the river fishing sites they have access to. This situation already takes place during the dry season. The perception & experience value will not change.

The discussed well-being impacts of the three types of flow regime changes for the identified stakeholder groups are summarised in Table 4.25.

Table 4.25 Summary of qualitative impacts on stakeholders resulting from three types of flow regime changes

Stakeholder group	% of population	Tipaimukh reservoir			Cachar plain irrigation			Morphological changes		
		I&F	H	P&E	I&F	H	P&E	I&F	H	P&E
Upstream										
• Farmers	48	0/-	+/0	+	-	+/-	+	0	-	0
• Fishermen	4	-	+/0	+	--	+/-	+	0/-	-	0
• Others	29	0	+/0	+	0	+/-	+	0	-	0
Downstream										
• Farmers	6	0/-	+/0	+	-	+/0/-	+	0	0/-	0
• Fishermen	1	-	+/0	+	--	+/0/-	+	0/-	0/-	0
• Others	12	0	+/0	+	0	+/0/-	+	0	0/-	0

When flooding reduces at the benefit of dry season flows, many stakeholder groups experience positive and negative impacts at the same time. Because people are poor, income and food may have the highest priority. Following the results in Table 4.25, this means that the people hit hardest are the fishermen, and within the fishermen group the poorest households. Negative impacts are largest when water is withdrawn upstream to irrigate the Cachar plain. However, no quantitative assessment has been made of the frequency, timing and extent of floodplain inundation under these different scenarios. Therefore, actual impacts remain uncertain. This section has mainly tried to provide an overview of what the possible impacts are and who are impacted in what way and to what extent. When additional information on the scenarios becomes available, the table provides the basic structure and can be updated.

4.8.3 Conclusions on combining all links

This section combined all links to understand the relationship between the upstream flow regime and human well-being. The data did not allow for a quantitative assessment, but were suitable to quantitatively evaluate possible impacts of flow regime changes. When low flows decrease, the poorest people who depend on the river for domestic use will be affected. Reduced peak flows may take away the flood pulse required to sustain the life cycle of fisheries and possibly also reduces opportunities for floodplain cultivation. Within the groups of fishermen and farmers, it will be the poorest households for whom such changes will have the severest impacts. All households, and again especially the poor, can be expected to be negatively affected by a reduced availability of locally produced food.

This information, was generated with the aim of providing input to negotiations with India on use of the Barak water. For such negotiations quantitative impacts of various water resources management alternatives for the Barak river will be required. Because of the limited availability of data, the analysis in this case study was qualitative. The described analysis has shown how human well-being impacts of flow regime changes can be assessed. With additional data, the results of this analysis could be quantified.

4.9 Discussion and conclusions

This chapter discussed an application of the conceptual model described in Chapter 3 in a case study in Bangladesh. In this section discusses the results of this application at three levels: 1) human well-being aspects of environmental flows for the Surma-Kushiyara rivers and floodplain, 2) suitability of the methods used to assess environmental flows from a well-being perspective, and 3) the usefulness of the conceptual model for the main topic of this dissertation: assessing human well-being values of environmental flows.

4.9.1 Human well-being aspects of environmental flow requirements for the Surma-Kushiyara floodplain

Three main types of stakeholders can be identified in the Surma-Kushiyara floodplain with respect to use of the river ecosystem: farmers, fishermen and others. Farmers and fishermen depend to a large extent for their income and food on the goods and services of the river and floodplain. A small part, approximately 10% of the inhabitants makes use of the river water for domestic purposes. The people depending on the river for domestic purposes are generally the poorest people: the fishermen and the landless labourers. The main link between perception & experience and the river ecosystem are inconveniences related to flooding of the settlement area. Again, the same poor people living in the worst areas are most affected by this flooding.

A qualitative analysis was made in which three types of changes were considered: 1) construction of the Tipaimukh dam for hydropower only, 2) use of Tipaimukh reservoir water for irrigation of the Cachar plain, and 3) further morphological changes leading to reduced low flows in the Surma. Especially fishermen will loose income when river and floodplains no longer connect at the right times. At the other hand, the fishermen and other poor households' health will benefit from the increase of river flows during the dry season. It is not clear whether the combination of hydropower and irrigation will lead to a net increase or reduction of dry season flows. All impacts are estimations and should be considered as input for further investigation and updating.

In this case study, flow requirements were based on the current use of ecosystem goods and services and the current flow regime. This holds the implicit presumption that the current situation is desired. However, there are reasons to doubt this presumption: use of ecosystem goods and services is adapted to the system of annual flooding. A change in flow regime will require a change in income-generating activities and other ecosystem uses, which may either improve or worsen the well-being of the population. The villagers complain about the negative impacts of flooding, and claim that precipitation alone will suffice crop water requirements. Possible adaptations to new circumstances are a topic for further investigation.

Additional data collection is required especially for the relationship between the river and floodplain ecosystem and the local flow regime, and for the relationship between the local flow regime and the upstream flow regime. For the relationship between the ecosystem and the local flow regime additional data collection and analysis have been done by Bari and Marchand (2006). The results from their study were not available at the time the analysis for this chapter was made. The analysis will greatly benefit from additional information on the upstream flow regime and how alternative operation strategies for the Tipaimukh dam and further morphological changes at the Surma-Kushiyara bifurcation point at Amalshid will change the flow regime in the Surma and the Kushiyara.

In Integrated Water Resources Management positive and negative impacts of interventions are preferably evaluated at the basin scale. In transboundary river basins, not only the basin scale is important, but also the impacts for each country. This information can be used in international negotiations. In these discussions, countries will tend to focus for their own country on the negative impacts and for the other countries on the positive impacts that can be expected in these countries. This will mean that for its negotiations with India, Bangladesh will naturally be more interested in the negative impacts of dam construction. Because of the assumption in environmental flow assessments that all changes in the flow regime will result in changes in the river ecosystem, environmental flow assessments are suitable to assess such negative impacts. However, for the local population, changing flow regimes may also have positive impacts, such as less inundation of the villages and an increase of dry season flows for domestic use. Although these flows should not be considered environmental flows, all impacts of changing flow regimes need to be considered in order to understand the importance of environmental flows. To consider social equity, it is important in all these evaluations at the various scales to not only consider the net benefit for a basin or a country, but to take into account *who* benefits and *who* looses. Winners and losers are often different groups of people.

4.9.2 Suitability of the methods used

The main data collection method used in the case study was interviews at the household level. Interviews with the population are useful to understand which of many potential ecosystem goods and services are actually used, and how important these are for the well-being of the people. Whether the population can also help to provide insight in the relationship between ecosystem goods and services and river flows will depend on the variation in flow regimes they have experienced. This can be the case when there is large natural variation or when interventions to the river system are already carried out. This was not the case in the Surma-Kushiyara system where recurring floods form a large problem, and people have presumably not often experienced the effect of low flows or of late floods.

In this case study the focus was on the well-being of the people, and their perception of the importance of the river ecosystem and the flow regime. Such a study should preferably be a component of a multi-disciplinary environmental flow assessment or water resources management analysis. Factual information on ecosystem condition and the river flow regime over a long range of years (more than 20 years), will be highly useful to place the current condition of the system in a larger context, which will benefit the formulation of questions and the understanding of answers in the interviews. Moreover, data on ecosystem condition and flow regime will allow for making quantified links between flows, ecosystem goods and services and human well-being, and subsequently for the quantification of the impacts of flow regime changes.

4.9.3 Usefulness of the conceptual model

The conceptual model, which was at the same time being applied and further developed based on the experiences in this case study in Bangladesh, helped to design the case study, and especially the structuring of the wide-ranging information collected in the interviews. During this structuring of interview information, the context was identified as key to understanding the importance of the river ecosystem for the various groups of people, and was added as a component of the conceptual model. The resulting conceptual model encourages explicit attention for the context, and allows for including the context in identifying realistic links between ecosystem condition and human well-being.

5 Well-being impacts of changes in hydrology of the Hamoun wetlands, Iran

5.1 Introduction

The Hamoun wetlands in the Iranian Sistan-Baluchestan Province, near the Afghan border, suffered from a severe drought between 1997 and 2005. In the wetland area, where people used to catch fish and harvest reeds, an expansive desert emerged. A large part of the population of the Sistan area depended mainly on the reeds, fish and birds provided by the wetlands. The wetland also played a role in regulation of the local climate. This chapter describes a second application of the conceptual model for the assessment of human well-being impacts of changed hydrology of the Hamoun wetlands resulting from alternative water resources management strategies for the Sistan area. First, the conceptual model is applied to assess quantified relationships between human well-being, the Hamoun ecosystem and the hydrology of the Hamouns. Second, the relationships are used to assess the impacts of changes in hydrology of the Hamoun wetlands as a result of various water resources management strategies following an IWRM approach.

The background of the Sistan area and objective for this case study are described in Section 5.2. Section 5.3 explains the approach and data collection methods used for applying the conceptual model. Stakeholder groups are identified in section 5.4. Section 5.5 discusses the relationship between the well-being of each of the stakeholder groups and the condition of the ecosystem, while the relationship between the condition of the wetland ecosystem and the hydrological regime of the wetland is the topic of section 5.6. Section 5.7 describes an application of the RIBASIM water balance model, with which inflows into the area are linked to the wetland hydrology. All relationships together are used in section 5.8 to estimate impacts of water resources management strategies on human well-being in the Sistan closed inland delta. Section 5.9 draws conclusions with respect to 1) human well-being impacts of changes in the Hamoun wetlands resulting from water resources management strategies, 2) the suitability of the methods used for applying the conceptual model and stepwise approach and 3) the usefulness of the conceptual model and the approach for the quantification of human well-being impacts related to environmental flows in an IWRM study.

5.2 Background and objective of the case Study

5.2.1 Description of the area

The Sistan closed inland delta in the east of Iran (Figure 5.1), suffers from regular droughts, which seem to be more severe in recent years. Ninety-five percent of the 150,000 km² large catchment is located in Afghanistan. Because of the low local precipitation of 60 mm/year, the Sistan area depends largely on Afghanistan for its water supply.

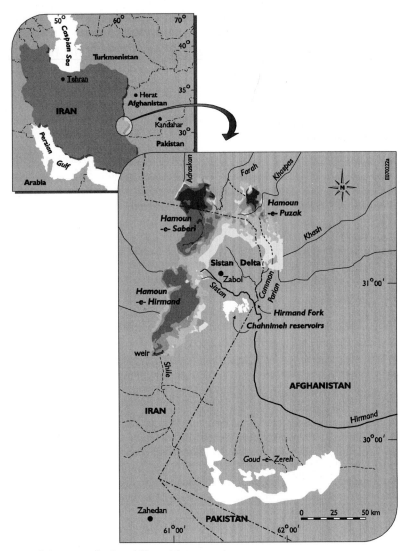

Figure 5.1 Map of Hamoun wetlands and Sistan delta

The main volume of water reaches Iran through the Hirmand river, ca. 5000 MCM/year, while other rivers contribute together another 2500 MCM/year. The Hirmand river bifurcates at Hirmand Fork into the Common Parian river and the Sistan river. Water from the Sistan river is diverted into the Chahnimeh reservoirs, with a total capacity of 800 MCM, to secure public water supply for the 400,000 inhabitants of Sistan delta and of two neighbouring cities. Further downstream, two weirs divert water for irrigation. Along the Common Parian river three inlets for irrigation canals exist. The total irrigated area of the Sistan area has a size of 120,000 ha. The water which is left after these abstractions flows into the Hamoun wetlands system. These wetlands consist of three distinct, but linked, parts: Hamoun-e-Puzak, Hamoun-e-Saberi and Hamoun-e-Hirmand. When full, the three Hamoun wetlands form one large wetland of 400,000 ha with an average depth of 2 m. During large floods, the wetlands spill into the Shile river, which flows into the Goud-e-Zereh lake across the border in Afghanistan. This salt lake is the actual end of the system. Regular spilling prevents the Hamoun wetlands from becoming saline as well.

During the 1970s the Hamoun wetlands contained vast areas of reeds, numerous species of birds and large amounts of fish. Moreover, winds blowing over the lake functioned as a regulator of the otherwise very hot and dry climate (Mansoori, 1994). During droughts, these winds resulted in sandstorms. Because of its biodiversity conservation value, the Hamoun wetlands were designated a Ramsar site in 1975. Since 1975, water use has intensified: the population has increased from 174,000 to 400,000 people, the irrigated area has increased from around 86,000 ha to the current 120,000, and the Chahnimeh reservoirs have been constructed. The registering of the wetland on the Montreux list in the 1990s, a list of threatened Ramsar sites, indicates that degradation was taking place at that time already. When in 1997 a severe drought of several years started, the area turned into a desert. Not only did this mean a loss of an ecologically valuable area, but also did it affect the life of the people depending on the wetland for various components of their well-being. The dependence of various groups of people on the wetland and other income-generating activities is central in this case study and will be discussed in more detail later in this chapter.

Water resources management in Iran is organised per sector. The main responsibility for large water resources infrastructure lies with the Ministry of Energy. The Ministry of Jehad-Agriculture is responsible for the supply from the main irrigation canals to farm plots and drainage into the main draining canals. From there, again the Ministry of Energy is responsible. The decision-making power for Sistan is organised at the *Ostan* (province) level, and is located in Zahedan, the capital of Sistan-Baluchestan Ostan. The Sistan closed inland delta is almost entirely located within Zabol *Shahrestan*, a subprovince of Sistan-Baluchestan. The national Ministries have regional authorities in Zahedan and local bureaus in Zabol. The representation of the Ministry of Jehad-Agriculture at the local level is done by several separate bureaus: the Agricultural Bureau, the Natural Resources Bureau, the Fisheries Bureau, and the Nomads Bureau. Also the Ministry of Energy has a number of local offices: the Zabol Water Board, the Rural Water and Waste Water Authority, the Urban Water and Waste water authority, and the Soil and Water Development company. Other relevant government authorities are the Department of the Environment, responsible for protecting the natural environment, and the Management and Planning Organisation responsible for allocating budgets to development projects.

Farmers, fishermen and nomads are represented by separate bureaus of the Ministry of Jehad-Agriculture. Although the Department of the Environment is responsible for the protection of the natural ecosystem, it does not represent the interests of ecosystem users, such as fishermen and bird catchers.

5.2.2 Problem analysis and objectives

Currently, the Hamoun wetlands, which are at the end of the chain, receive the water which is left are filling the Chahnimeh reservoirs and supplying irrigation demands. Since the start of the drought in 1997, this amount has been very low, with negative consequences for the people depending on the wetland for their income and for the entire Sistan population because of the aggravation of sandstorms. In an IWRM study, initiated by the Ministry of Energy of Iran, the impact of water resources management on the socio-economic situation of the area was analysed (see for full report on the IWRM study Van Beek & Meijer, 2006). Irrigated agriculture is important for income-generation, but so are wetland uses such as bird-hunting, fisheries, and livestock-herding. Industry is of minor importance in the area. The challenge of the IWRM study was to present an unbiased view of the economic and social benefits and losses of different distributions of water between agriculture and natural resources.

For the water resources management of Sistan, the following objective was formulated in the IWRM study:

> *To support the socio-economic development of the Sistan area on the basis of sustainable resource use (surface water and groundwater), while protecting and restoring the natural environment.*

The two main results of the study are: 1) a flow-forecasting system to assess the availability of water, and 2) an IWRM analysis describing the impacts of strategies to deal with the forecasted flow. Forecasting of flows through remote sensing data and meteorological forecasts is required, because the availability of hydrological data of the Afghan part of the basin is highly limited. One component of the IWRM analysis is the analysis of impacts of water resources management strategies on the wetland ecosystem, and subsequently on the well-being of the Sistan population which benefits from this wetland ecosystem in various ways. The case study discussed in this chapter was carried out to fulfil this part of the IWRM analysis as well as to test the conceptual model and approach developed in Chapter 3. The specific objective for this case study is:

> *To quantify for various water resources management strategies the well-being impacts for the population of Sistan resulting from changes in wetland hydrology and ecosystem condition.*

5.3 Method: approach & data collection

5.3.1 General approach

To quantify well-being impacts through an application of the conceptual model, the five steps, as discussed in Chapter 3, were translated to the situation of this case study:

1. identifying stakeholder groups;
2. assessing the relationship between well-being and the Hamoun wetlands ecosystem;
3. assessing the relationship between the Hamoun wetlands ecosystem and the Hamoun hydrology;

4. assessing the relationship between the Hamoun wetland hydrology and the upstream flow regime; and
5. estimating the impacts of changed Hamoun hydrology resulting from water resources management strategies on the well-being of the identified stakeholder groups.

Step 1. Identifying stakeholders groups

Stakeholders, defined as all people whose well-being is affected by changes in the condition of the ecosystem, were identified through an inventory of potential wetland goods and services, and of people who could possibly benefit from these goods and services. This was done based on literature, on interviews with local authorities in Zabol, the area's main city, and on group discussions with local communities. The results of this step are discussed in section 5.4.

Step 2. Relationship between well-being and the Hamoun wetlands ecosystem

Group discussions with groups of stakeholders provided initial information about stakeholder groups and the well-being values to which the Hamoun wetlands contribute. Based on the information from the group discussions, for each well-being value, indicators and qualitative relationships between indicators and ecosystem condition were identified. To quantify the relationship between indicators and ecosystem condition, a questionnaire survey was conducted. The survey results were analysed using simple statistics: either by counting the number of people who have given a certain answer, or by averaging the answers. This was done separately for each of the identified stakeholder groups. The results of this step are discussed in section 5.5.

Step 3. Relationship between the Hamoun wetlands ecosystem and the Hamoun hydrology

To establish a relationship between the hydrological regime and the ecosystem condition of the Hamoun wetlands, an approach based on the principles of DRIFT was applied. In this approach, hydrological parameters and ecological condition were assessed for a reference situation. In this study, relevant hydrological parameters were identified through analysis of data on flooding extent, spills, vegetation cover and numbers of birds. Thresholds were identified above or below which a significant ecosystem change was expected. Then, for all combinations of parameters and threshold levels, the associated ecosystem condition was estimated as percentage of the reference condition. The results of this step are discussed in section 5.6.

Step 4. Relationship between the Hamoun wetland hydrology and the upstream flow regime

The Hamouns are fed through various streams from which water is withdrawn for a number of purposes. To assess the relationships between the hydrological regime of the wetland and the upstream flow regime in various locations, use was made of an application of the water balance model RIBASIM[3] for Sistan, developed in the IWRM study. In this water balance model, water supply, water use, and management rules are combined. Step 3 revealed that the most relevant hydrological processes took place at

---

[3] River Basin Simulation software developed by WL | Delft Hydraulics for analyses of water resources systems.

the time-scale of seasons and years. Therefore, a simulation time-step of one month was considered suitable for obtaining the required information for calculation of the hydrological parameters. The results of this step are discussed in section 5.7.

Step 5. Impacts of changed Hamoun hydrology resulting from water resources management strategies on the well-being of the identified stakeholder groups

With the established relationships, the well-being impacts of changes in Hamoun hydrology were assessed. The investigated changes in hydrology are the estimated results of alternative water resources management strategies. The strategies were simulated with the RIBASIM model. Next, the results from RIBASIM were translated into values for the identified hydrological parameters. Combining this with the hydrology-ecosystem relationship gave the average ecological situation resulting from that strategy, as well as the number of years during which the ecosystem is expected to be in a certain state. The change in ecosystem condition compared to the reference situation was then used to calculate for each stakeholder group the change in well-being in terms of the identified indicators. The human well-being impacts of all strategies are presented together in a scorecard. This information will provide input to the Integrated Water Resources Management process of decision-making. The results of this step are discussed in section 5.8.

5.3.2 Data collection

Data collection for steps 1 and 2 consisted of:

- interviews with authorities;
- group discussions with local communities;
- survey among local population.

Data collection for step 3 and step 4 can be found in Penning and Beintema (2006) and Meijer and Van Beek (2006) respectively. Step 5 required no separate data collection, but combined the information of all other steps. The question lists of the interviews, group discussions and survey can be found in Meijer and Van Beek (2006)

Interviews with local authorities in Zabol

Local authorities represent specific groups of the local population. Therefore, interviews at the local authorities were the first step to identify groups of people to whom the wetlands are important. General information on the responsibilities and perception of the water resources management situation was obtained as well. The following authorities were consulted:

- Agricultural Bureau, Zabol;
- Fisheries Bureau, Zabol;
- Nomads Bureau, Zabol;
- Natural Resources Bureau, Zabol.

Group discussions with identified Hamoun user groups

To obtain a first understanding of the situation with respect to hydrology, ecosystem condition and values for well-being, group discussions were conducted with the groups of Hamoun users that could be identified based on the interviews with authorities and literature. The group discussions had the following objectives:

- assessing the values of the wetland for the local population in a qualitative way;
- assessing hydrological features of the area understood by the population;
- assessing how changes in the hydrological features influence the wetland values;
- gaining an understanding of the general importance of the Hamoun wetland compared to other aspects of peoples' well-being; and
- identifying additional stakeholder groups or further sub-divisions in stakeholder groups.

Five discussions have been conducted. Four male groups, with nomads (pastoralists), bird catchers, farmers and fishermen. One group discussion was conducted with women, who belonged to nomads and fishermen households. The number of participants varied between 2 and 15 participants. An Iranian student conducted the group discussions and translated the results.

Questionnaire Survey

To obtain more detailed information on the importance of ecosystem goods and services for the identified stakeholder groups, a questionnaire survey was designed and conducted. In this questionnaire, the use and values of the ecosystem were discussed for three situations: 1) the natural situation before the drought, 2) the start of the drought and 3) the situation after 7 years of drought (2004). For certain questions only one or two of these situations were used.

The entire population of Sistan (estimated at 71,200 households) benefits from the Hamoun wetlands and hence can be considered a stakeholder. A distinction can be made between people who use the Hamoun wetlands for income-generation, people who make use of other water-related activities for income-generation, and others. Further specification of these three groups led to a division of the Sistan population into eight stakeholder groups. Table 5.1 lists the identified groups with the estimated number of households for 1975 and 2005: five groups of Hamoun-users for income, two groups of water users for income (farmers) and one group of not water users for income (urban people). Because the focus of the study was on the Sistan delta, possible stakeholders outside this area were not considered in the analysis. Section 5.4 discusses the identification of the stakeholder groups and the estimation of the number of households in each group in more detail.

Contrary to what the identification of stakeholders suggests, people normally carry out various activities. People are assigned to one of the 8 groups based on their main income-generating activity in the situation before the start of the drought (i.e. the largest contribution to their total income indicated by the stakeholders themselves). Another important characteristic of the people, which may result in a different dependency on or perception of the Hamoun wetland is age. Three age groups were distinguished: 25 and younger, between 26 and 45, and older than 45.

Table 5.1 Identified stakeholder groups and number of conducted interviews

Stakeholder groups		Number of households		Age up to 25		Age 26 to 45		Age above 45		Total
		1975	2005	Male	Female	Male	Female	Male	Female	
Hamoun users	Bird catchers	570	1,000	9	6	9	2	13	-	39
	Fishermen	1400	2,500	14	1	9	4	11	2	41
	Reed harvesters	290	500	3	5	2	10	3	5	28
	Pastoralists	6,300	11,000	6	7	6	6	9	3	37
	Field cultivators near Hamoun	4,300	3,000	9	4	9	3	11	2	38
water users	Animal farmers	1,700	7,500	2	-	-	1	5	-	8
	Field cultivators far from Hamoun	10,800	18,900	1	-	12	1	12	-	26
other	Urban people	5,000	26,800	5	-	14	4	7	1	31
	Total	30,360	71,200							248

The interviews were conducted by students from Zabol University. The interviewers were informed through documents and discussion on the purpose of the questions, way of asking the questions, and were supported during their work. The students entered the results in a database following a pre-defined coding system. In total 248 interviews were conducted.

5.3.3 Selection of participants for group discussions and questionnaires

Group discussions were organised through contacting one person in the target group, who was asked to invite another 5 to 10 persons. The persons first contacted were selected by a local retired wetland ranger. This was considered the best approach, because Zabol Shahrestan is reportedly not a safe area. The poor socio-economic condition and the location near two international borders (Afghanistan and Pakistan) have resulted in smuggling and associated unsafety. Foreigners as well as local town's persons are treated in the rural areas with suspicion. When group discussions and interviews are organised through trusted contacts, the chance of cooperation and trustworthy answers may be higher. This was considered more important than the possible bias resulting from non-random selection.

For the interviews with the five Hamoun user groups villages surrounding the Hamouns were selected. The interviews with the animal farmers and field cultivators farther from the Hamouns took place near Hirmand Fork, and in Zabol City the interviews with urban people were conducted.

5.3.4 Limitations

Due to logistical constraints, the questionnaire survey had to be supervised from a distance. Although interviewers were well instructed, it is possible that some of the questions were misunderstood. If this could be assumed from the answers, the questions and answers were not considered in the analysis.

In the Iranian Muslim culture it is difficult to have women interviewed, especially by men. At the same time it is hard to find female interviewers. As a result most of the respondents were men. Although the participants were asked to answer not only for themselves but for the entire household, there may be a gender bias in the answers.

The focus in this thesis is on the human well-being – ecosystem condition links. To assess the human well-being impacts of alternative water resources management strategies, in this case study the entire chain of links from human well-being through ecosystem condition to the river flow regime is assessed. For the links between ecosystem condition, wetland hydrology and upstream flow regime use was made of work done in the framework of the IWRM study.

Especially for the link between hydrology of the wetlands and condition of the ecosystem very limited data were available. The links used in this case study were based on assumptions and expert judgement. Establishing and using these links in the framework of the IWRM study was highly useful to understand the data requirements, the connection with the other links and the types of results that can be obtained. This knowledge will be the basis for further investigations.

5.4 Identifying stakeholder groups

5.4.1 Wetland functions and potential stakeholders

In this case study, it is assessed what the difference in effect for different groups of people is, as a result of changes in the Hamoun wetland hydrology, which result from alternative water resources management strategies. Therefore, it is at first important to identify these different stakeholder groups. Stakeholders are all people whose well-being may be affected by changes in the flow regime. Through an analysis of the possible functions of the wetland an overview is made of what people are affected for what component of their well-being.

Because the wetland was during the study in a degraded state[4], to get a good impression of the goods and services the wetland provided in a healthy state, a reference situation should be considered. As reference situation the 1970s are chosen. For this period it is known that the wetlands were healthy. In the 1970s the wetlands were surrounded and partly covered with reed beds. Satellite images show a vegetation cover of 160,000 ha (in 1977). Large groups of migratory birds frequented the area in winter, and there were large groups of residential birds as well. Mansoori (1994) mentions the presence of 190 species of birds, with total numbers up to 700.000 in 1977. The Hamoun wetlands were an important source of fish. Numbers of fish are not known exactly. The Fisheries Bureau in Zabol mentions annual catches of 12,000 ton fish, while a leaflet of the Agricultural Bureau (Meijer, 2006) speaks of

---

[4] After spring floods in 2005 the wetland ecosystem partly recovered.

9,000 ton/year. The area harboured rare and threatened endemic species of fish and birds. Some reptiles and mammals were found as well.

In 1975, the Sistan population counted approximately 30,000 households, of which 25,000 households were residing in rural and 5,000 households in urban areas (Iran Statistics Institute, 1976). For a large part of the rural population the Hamoun wetlands provided livelihoods. However, the Hamoun wetland was not only important for income-generating activities. In the group discussions, the former Hamoun-users described life with a healthy wetland system as 'the wetlands provided everything': the climate was pleasant and the health of the people was good, despite the fact that they drank water from the Hamouns. Also the family relations were different: the nomads all lived together and shared their belongings: 'there was no separate fridge'. Moreover, the Hamoun wetlands were the scene for festivities and ceremonies. Summarising, the general picture that was obtained from the group discussion is that although the Hamoun users were not rich, they seemed in disposal of their direct needs and were leading a satisfied or even happy life.

In the 1970s the area of the Hamoun wetlands would decrease during summer, but normally the wetlands did not dry out completely. Because of the good condition of the ecosystem, the people were able to obtain sufficient food and income during the wetter times of the year to be able to cope with the reduced availability of wetland resources in the dry months. Also in those periods severe droughts occurred, during which the wetlands were completely dry for part of the year. This happened in 1970, as the villagers recall. At that time there were no alternative sources of water, not even for domestic use. As a result, the drought forced people to migrate.

As was mentioned before, after the 1970s various changes have taken place: a doubling of the population between 1976 (174,000 people) and 2005 (404,000 people) (Iran Statistics Institute, 1996), the construction of the Chahnimeh reservoirs (1981) and extension and modernisation of the agricultural area (construction of Shib-Ab and Posht-Ab irrigation areas around 1983). Also in Afghanistan developments may have taken place, although it is possible that due to the political problems many projects have not been carried out as planned.

The analysis of functions of the Hamoun wetland is done by applying the functions classification by De Groot (1992b, see also section 3.4) to the reference situation. The inventory below is only an inventory of observed functions, several of these functions may turn out to be easily replaced by alternatives, or not be given high priority by the population. In Section 5.5 the importance of the goods and services for the well-being of the people of the Sistan delta will be discussed in detail.

Carrying functions
When the Hamouns are filled with water, fishermen travel by boat to other villages to buy wheat and other products. This function can contribute to various values: income and food, through access to the market, and to perception & experience (own transportation and contacts with others).

Natural production function
The Hamoun wetlands have a very important natural production function. The main products which are harvested are fish, birds, and reeds. Various people obtain their

food and income through catching or harvesting these products. The products are also important at the local market; they provide fresh, varied and affordable food for the entire population of the Sistan delta.

Joint production function

The pastures surrounding the Hamoun lakes, which require regular flooding, are important for livestock breeding activities. There are nomads in the area, who travel around with their herds, but also settled farmers who let their cattle graze on these pastures. There is no large scale recession agriculture, but the former pastoralists mention that they grew fruits and vegetables on the land that had been inundated. Due to the fertile soil these crops were easily grown. Like the natural production functions, the joint production function provides income for the livestock herding people, but also ensures the availability of meat and dairy products at the local market, benefiting the entire population of the Sistan delta.

Signification functions

For the people living close to the Hamoun wetlands, the wetlands were the scene for traditional ceremonies like weddings and other parties. People further away visit the area for recreation, to enjoy. The ancient buildings on the Kuh-e-Khajeh, an old volcano, are the aim of their trip, but the beauty of this site lies in the surrounding of the mountain with reeds and water.

The wetland has not only a local, but also a global function. Being on the north-south route of migrating birds, it plays a role in sustaining global biodiversity. This intrinsic value is appreciated by nature lovers throughout the globe. The importance of the site is reflected in the designation as 'Ramsar'-site, although, as was mentioned already, the recent degradation has put the wetland on the Montreux-list, the list indicating threatened wetlands.

Processing functions

It is not clear to what extent this function plays a role in case of the Hamoun wetland. Part of the waste water from irrigation is collected in separate drainage basins, isolated from the wetland through dikes. Waste water from households is infiltrated in the soil (rural areas) or treated (urban areas) and reused. If the wetlands would play a role in waste processing, the people benefiting most from this would be the people in the direct proximity or downstream of the wetland. During the drought, waste accumulation has not turned out to be a problem.

Regulation functions

A very important regulation function of the Hamoun wetland is the regulation of the local climate and the prevention of sandstorms. The notorious 120-days wind blows around the months of May, June, July and August at a wind force of 4-6 Beaufort. During these months the temperature may reach above 40^0C. When the wetland is inundated during this period, these winds pick up and spread small water droplets over the area, this way relieving the heat. However, when the lakes are dry, the winds bring nothing but sand and dust. The resulting sandstorms have reportedly covered complete houses. Also numerous people suffer from skin, lung and eye complaints as a result of the dust. Women suffer even more because the dusty circumstances require frequent cleaning.

5.4.2 Stakeholder identification & numbers

Stakeholder groups can be distinguished based on different relationships between goods and services of the Hamoun wetland and human well-being values. Assumed is that for the groups of people who use the Hamouns for income, the Hamoun will be of high importance. Probably, those people use the Hamouns for other well-being components as well. The people who do not use the Hamouns for income, may still find the Hamouns important because of contributions to health, and possibilities for recreation. A distinction is made between people who depend for income on other water related activities (agriculture), and people who don't. Assumed is that since the Hamouns compete with agriculture, farmers may appreciate the Hamoun services less than the urban people may do. Stakeholders of the Hamoun wetlands are not all located in Iran; the Hamoun wetlands extend into Afghanistan, where there will be Hamoun users for income as well. The prevention of sandstorms will be beneficial to people in Afghanistan and Pakistan, while the sustenance of global biodiversity may be appreciated worldwide. Because the focus of the study has been on how water resources management in Iran can improve the situation for the local people, those foreign stakeholders are not further considered. To be able to investigate the differences between groups, and thus social equity, the groups of Hamoun users for income and of farmers are further divided in bird catchers, fishermen, reed harvesters, pastoralists and field cultivators (near Hamouns), and in people in animal husbandry on farm and field cultivators (far from Hamouns). Figure 5.2 summarises the derivation of stakeholder groups.

The number of households in each group has been estimated based on population statistics for rural and urban areas, and on discussion with people at authorities and in the field. At first an estimation of the total population in 2005 in Zabol Shahrestan was made based on growth rates derived from population statistics (Table 5.2). In Iran, a population census is conducted every ten years, the last census was in 1996. The population in 2005 was estimated based on the growth rate over the period 1986-1996. Separate growth rates were assessed for the urban (3.78%) and rural population (1.24%), to take urbanisation into account. Household numbers were estimated based on the household size (in 2002) for urban areas of 5.67 persons, and for rural areas 5.68 (Iran Statistics Institute, 2002). This leads to a total of 26,800 urban and 44,400 rural households in 2005.

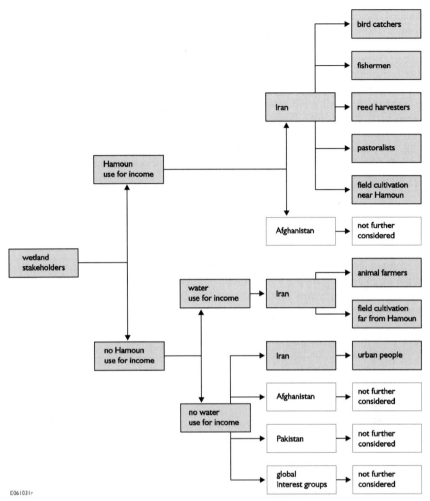

Figure 5.2 Identification of stakeholder groups

Table 5.2 Population numbers of Zabol Shahrestan for 1966-1996, with estimation of the population and households in 2005 (Source: Iran Statistics Institute, 1966; Iran Statistics Institute, 1976; Iran Statistics Institute, 1986; Iran Statistics Institute, 1996).

Year	1966	1976	1986	1996	2005	Households 2005
Urban population	18,806	29,404	75,105	108,889	152,114	26,828
Rural population	141,731	144,898	199,506	225,672	252,143	44,391
Total	160,537	174,302	274,611	334,561	404,257	71,219

For the analysis in this chapter, it is required to understand of how many households each of the eight stakeholder groups consists. Assumed is that all 26,800 urban households have not water-related activities as their main activity. The distribution of the 44,400 rural households over the seven remaining stakeholder groups was estimated based on various sources of data. The Agricultural Bureau of Zabol mentions about 30,000 cattle herding families, 15,000 of them on the borders of the Hamouns (referred to here as pastoralists), and 15,000 on farms (referred to as animal farmers) (Penning & Beintema, 2006). Another group of pastoralists are the nomads, 3200 households are under the cover of the Nomads Bureau. According to the Fisheries Bureau, the number of 2500 fishermen's households in 1999 has not increased since. In all group discussions the number of households depending mainly on reed harvesting is mentioned to be very small. Bird catchers mentioned during the group discussion that in and near their city 300-400 bird catching households were living, and this was similar in a few other places. A total of 1000 households is assumed. Most of the rural households involve in field cultivation. Their number is calculated by abstracting above numbers from the total number of 44,400 rural households. Of the resulting 21,900 households, 3000 are assumed to live near the borders of the Hamouns, and 18,900 further away. During the group discussion and interviews it has become clear that most of the households involve in more than one activity. Yet, they refer to themselves as belonging to a certain group based on what they consider their main activity. The numbers, which are summarised in Table 5.3, will contain uncertainties, but are expected to give a good representation of the relative size of each group within the community of Zabol Shahrestan. For 1975 the number of households was estimated using the population numbers of 1975 and the distribution over activity groups of 2005.

Table 5.3 Number of households in the identified stakeholder groups

	Stakeholder group	Households 1975	Households 2005
1	Bird catchers	600	1,000
2	Fishermen	1,400	2,500
3	Reed harvesters	300	500
4	Pastoralists	6,300	11,000
5	Field cultivators near Hamoun	1,700	3,000
6	Animal farmers	4,300	7,500
7	Field cultivators far from Hamoun	10,800	18,900
	Total rural households	25,500	44,400
8	Urban people	5,000	26,800
	Total households	30,400	71,200

According to these numbers, approximately 25% of the population uses the Hamoun wetlands for income generation, while 37% makes use of water for other income-generation purposes and another 38% does not depend on water for their income. However, as was mentioned before, many people carry out various activities for income-generation.

5.5 Relationship between human well-being and the Hamoun wetlands ecosystem

This section discusses the results of step 2: linking human well-being and the ecosystem condition. At first, the section discusses the contribution of the wetland to total human well-being, which is the top box of the conceptual model. Then, for each of the main well-being values, the link with the Hamoun wetland and the context are discussed. Subsequently, operational indicators are defined and the approach to calculate indicators scores is explained. If not mentioned otherwise, the results discussed are the results of the questionnaire survey.

5.5.1 Contribution of the Hamoun wetlands to total human well-being

Aggregating the values of the different components into one well-being value requires subjective weighing and will lead to a loss of transparency. Nevertheless, information on the general appreciation of the wetland provides insight in the importance of the wetland for local communities. Therefore, the importance of the wetland for the stakeholder groups in the situation before the drought was analysed. In this analysis, the context was not considered explicitly, but through the indication that the wetland is not important for certain components of well-being and by directly stating whether the wetland is perceived as valuable or not, the relative importance of the wetland compared to other sources which contribute to human well-being is revealed.

A first indication of the importance of the Hamoun wetlands is the number of people who indicate the use of the wetland for a certain purpose. Table 5.4 reveals that most of the people have the perception that the Hamoun wetlands contribute to various components of their well-being. In urban areas this perception is somewhat less, but the contributions to food, living condition, health and recreation are acknowledged as Hamoun wetland values by virtually all people of this group as well. Interestingly, also within the stakeholder groups living far from the Hamoun wetlands, or in the cities, many people indicate the use of the Hamoun wetlands for income-generation. Perhaps they obtain part of their income from the wetlands, or they value the wetland as a source of income for others. For each value one or more sub-components were used, following the description of goods and services, to avoid misunderstanding about what the main well-being values comprise (Table 5.4). The values and sub-components have some overlap on purpose, because people may not directly link, for example, domestic water and health.

Table 5.4 Percentage of participants stating contribution to various values by the Hamoun wetland (shaded areas mark values perceived by less than 80% of the participants in the stakeholder group)

values			Income & Food				Health			Perception & experience		
Stakeholder group	Nr. participants	% of population	Employment & income	Food	Materials	Livestock	Domestic water	Living conditions	Health	Recreation	Religion/ Traditions	Transport/ Communication
Bird catchers	39	1.4	100	100	92	59	100	100	100	100	85	87
Fishermen	41	3.5	100	100	98	73	98	100	100	100	90	100
Reed harvesters	28	0.7	100	100	100	71	100	100	100	100	96	93
Pastoralists	37	15.4	100	100	97	100	100	100	100	100	92	95
Field cultivators near Hamoun	38	4.2	92	100	87	95	92	100	100	95	84	79
Animal farmers	8	10.5	88	100	38	100	75	100	100	100	100	63
Field cult. far from Hamoun	26	26.5	92	100	12	92	85	96	100	100	54	15
Urban people	31	37.8	29	97	6	35	65	97	100	100	32	6

Since the start of the drought in 1997, the Sistan-Baluchestan Rural Water and Waste Water authority has invested in public water supply. According to this authority, since a few years, 100% of the villages, even single farms, has piped water from the Chahnimeh reservoir at its disposal (Meijer & Van Beek, 2006). Therefore, domestic water can nowadays be considered a value for the nomadic tribes only. This leaves the contribution to employment and income, food, living conditions and climate and health as most important values of the Hamoun wetland.

Although Table 5.4 shows that the wetland contributes to various well-being values, this does not yet imply that the wetland is indeed important compared to possible other contributions to the well-being of the stakeholder group. Therefore, the participants were asked in addition whether in general the Hamoun wetlands are considered important or not. Table 5.5 shows that the wetland is indeed important to virtually all people of Sistan delta. For the people who use the Hamoun wetland for income, the wetland is very important, while people in the cities and further away from the Hamouns classify the wetland simply as important.

Table 5.5 Percentage of stakeholder groups indicating the Hamoun wetland as not important, important or very important

	not important	important	very important
Bird catchers	0	5	95
Fishermen	0	0	100
Reed harvesters	0	14	86
Pastoralists	0	5	95
Field cultivators near Hamoun	0	24	76
Animal farmers	0	25	75
Field cultivators far from Hamoun	0	54	46
Urban people	0	61	39

From the findings presented in this section, it can be concluded that the Hamoun wetlands contribute to human well-being in various ways. Although the stakeholders who use the Hamoun wetland for income have perhaps the most direct relationship with the Hamoun wetland, virtually all people of the Sistan delta consider the Hamoun wetland important for some components of their well-being. The next sections will consider in more detail the relationship between the Hamoun wetland and the three main well-being values, for each of the stakeholder groups.

5.5.2 Income & food generation

Link with ecosystem

The wetland resources reeds, fish and birds are used for income-generating activities by the population. When resources are available in abundance, some degradation can take place without affecting the well-being of the people. This was probably the case in the reference situation (1970s). About this period bird catchers indicated in the group discussions that a larger number of birds would not have brought them a higher income, because it was simply not possible to catch more. However, in recent years, this abundance does no longer seem to be present. Penning and Beintema (2006) identify over-exploitation of the ecosystem as one of the reasons for degradation of the ecosystem. This means that each level of ecosystem degradation will result in a reduction of resource use. Therefore, a linear relationship is assumed between ecosystem condition and use of resources.

Context

The relationship between resources and income for the different stakeholder groups is further influenced by the following aspects of the context:

1. importance of Hamoun-related versus other activities;
2. availability of alternatives during drought situation;
3. access to goods and services;
4. preference for changes from the natural situation; and
5. over-exploitation.

Importance of Hamoun-related versus other activities

Although the stakeholders were assigned to an activity-related group, all people make use of various resources. An overview was made of the contribution of the various activities to the total income of each group. Table 5.6 shows that the stakeholder groups who have a Hamoun-use activity as their main activity (bird catchers, fishermen, reed harvesters and pastoralists), depend for around 90-95% on the Hamoun wetlands for their income. This puts high importance on the Hamoun wetland for the well-being of these groups. For the other four stakeholder groups, the Hamoun wetlands are less important for the generation of their income. The contribution of the various activities to total income will be used to calculate the change in income as a result of changes in the resources related to the activities.

Table 5.6 Contribution (in %) of various activities to total income of stakeholder groups before drought

Stakeholder group	Fishing (A)	Bird catching (B)	Reed harvesting (C)	Animal husbandry on pastures (D)	Field cultivation near Hamoun (E)	Animal husbandry on farm (F)	Field cultivation far from Hamoun (G)	Horticulture (H)	Other water-related activities (I)	Other activities (J)	Total direct Hamoun use (A+B+C+D)
Bird catchers	20.9	56.1	15.4	3.5	2.5	0.0	0.1	0.0	1.2	0.3	95.9
Fishermen	68.6	5.9	18.4	5.5	1.3	0.1	0.0	0.0	0.0	0.1	98.4
Reed harvesters	17.1	4.5	67.3	8.3	1.6	0.4	0.0	0.0	0.0	0.7	97.3
Pastoralists	1.0	0.4	3.6	82.2	11.0	1.8	0.0	0.1	0.0	0.0	87.1
Field cultivators near Hamoun	2.2	0.5	1.9	18.5	66.0	9.5	0.0	0.1	0.0	1.3	23.1
Animal farmers	1.5	0.0	0.0	8.4	11.6	55.6	22.9	0.0	0.0	0.0	9.9
Field cultivators far from Hamoun	0.0	0.0	0.0	1.2	0.0	24.8	71.7	0.8	0.0	1.5	1.2
Urban people	0.0	0.0	0.0	1.6	2.4	3.4	8.1	0.3	0.0	84.1	1.6

Availability of alternatives during droughts

During droughts, the people depending on water for their income, are forced to find alternative sources of income. Table 5.7 shows that during the drought, all stakeholder groups, except the group of animal farmers, obtain more than 80% of their income through other, not Hamoun or water-related, activities. However, this higher share of other activities does not yet imply that losses were sufficiently replaced. Of the Hamoun-using people (defined as the people who obtained before drought more than 50% of their income through Hamoun-using activities), 75-90% indicate that their income during drought is very small compared to the reference situation (see Table 5.8). Among field cultivators and animal farmers, this number is 50-80%. The urban people obtained during drought an income largely (60%) similar to their income before drought. This shows that the stakeholder groups who depend on the Hamouns or other water-related activities have not been able to replace the loss of their income due to drought. Moreover, more than 97% of the Hamoun-using people were willing to return to their original activities. Again, this shows that no satisfying alternatives were found by the population.

Table 5.7 Contribution (in %) of various activities to total income of stakeholder groups during drought

Stakeholder group	Fishing (A)	Bird catching (B)	Reed harvesting (C)	Animal husbandry on pastures (D)	Field cultivation near Hamoun (E)	Animal husbandry on farm (F)	Field cultivation far from Hamoun (G)	Greenhouse (H)	Other water-related activities (I)	Other activities (J)	Total direct Hamoun use (A+B+C+D)
Bird catchers	0.0	0.0	1.3	0.0	0.0	0.0	0.0	0.0	0.0	98.7	1.3
Fishermen	4.9	0.7	5.4	0.0	0.0	0.0	0.0	0.0	0.0	89.0	11.0
Reed harvesters	0.0	0.0	19.6	0.0	0.0	0.0	0.0	0.0	0.0	80.4	19.6
Pastoralists	0.0	0.0	3.8	2.4	0.0	4.3	0.0	3.2	0.0	86.2	6.2
Field cultivators near Hamoun	0.0	0.0	1.3	0.0	0.0	2.6	0.0	0.0	0.0	96.1	1.3
Animal farmers	0.0	0.0	0.0	6.3	0.0	27.5	3.8	0.0	0.0	62.5	6.3
Field cultivators far from Hamoun	0.0	0.0	0.0	0.0	0.0	9.4	6.7	0.0	0.0	83.8	0.0
Urban people	0.0	0.0	0.0	0.0	0.0	1.6	0.0	0.0	0.0	98.4	0.0

Table 5.8 Percentage of people in each stakeholder group who perceived their current income to be similar, half or very little of their income before the drought

Stakeholder group	Similar amount	About half of amount	Very little amount	Missing*
Bird catchers	2.6	20.5	74.4	2.6
Fishermen	2.4	7.3	90.2	0.0
Reed harvesters	3.6	10.7	82.1	3.6
Pastoralists	0.0	10.8	86.5	2.7
Field cultivators near Hamoun	0.0	21.1	78.9	0.0
Animal farmers	12.5	37.5	50.0	0.0
Field cultivators far from Hamoun	30.8	11.5	53.8	3.8
Urban people	58.1	32.3	9.7	0.0

* % of participants that did not give an indication for their change of income

From informal discussions, it is known that people have obtained alternative income through smuggling, mainly of gasoline. Because of the sensitivity of the topic, this could not be included in the group discussions and interviews. It is possible that through these illegal activities people obtain a higher income. To obtain insight in the preference of people, it was asked whether the people preferred to return to their previous activities if the ecosystem recovered. This was asked only to the people who obtained 50% or more of their total income before drought through the combined Hamoun-use activities (146 respondents). Table 5.9 shows that in all groups, virtually all people prefer to return to the previous activities. Of the two people who did not want to return to their pre-drought activity, one indicated that this was due to his old age. There are no indications that people prefer the alternative activities above the situation before the drought.

Table 5.9 Number of people hoping to return to Hamoun using activities if wetland is restored

Stakeholder group	Number of respondents	Yes	No
Bird catchers	39	97.4	2.6
Fishermen	41	100	0
Reed harvesters	28	100	0
Pastoralists	37	97.3	2.7
Field cultivators near Hamoun	1	100	0

Access to goods and services

Fishing and bird hunting are banned during spawning and breeding periods respectively. However, the discussions with the population did not give indications that certain areas or activities are restricted to certain groups of people only. Therefore, changes in availability of resources are assumed to affect all users in the same way.

Preference for changes from the natural situation

Although the Hamoun plays a large role in income-generation, it is possible that, if available, alternative income-generation activities would have been preferred. The question of which of four types of activities would be preferred if available: agriculture, Hamoun use, industries or services, was asked again to the people for whom 50% or more of their total income comes from the Hamoun wetland. The data show that many people are interested in shifting to industrial work, but that nevertheless Hamoun use is largely preferred (Table 5.10).

Table 5.10 Percentage of stakeholder groups with preference for Hamoun use, agriculture, industry and services

Stakeholder group	Number of respondents	Hamoun use	Agriculture	Industry	Services
Bird catchers	39	100	0	0	0
Fishermen	41	92.7	2.4	4.9	0
Reed harvesters	28	67.9	3.6	25.0	3.6
Pastoralists	37	54.1	13.5	16.2	16.2
Field cultivators near Hamoun	1	0	0	100	0

The younger generation has less experience with the situation before the drought. Therefore, they may be less interested in living from Hamoun-use in the way their parents and grand-parents did. Possibly, they are higher educated as well. For this reason, a second analysis was carried out based on a division according to age instead of activity group. All respondents whose income is for 50% or more dependent on the Hamoun wetlands were divided into three groups: young (<25 years), middle-aged (25-45 years) and old (>45 years). Because the number of people in each of the activity groups are not representative of their number in society, this different arrangement in age-groups is biased for activity. Therefore, the results of this analysis cannot be used to say something about the population as a whole, but only about the current selection. Because only the group of Hamoun-users are included, it still gives an impression of the perception of the situation with respect to income-generation before and during drought. Table 5.11 shows that people of all ages prefer to continue using the Hamoun wetlands. Old people have a slightly higher tendency to stick to what they used to do, between young and middle-aged people there are no significant differences.

Table 5.11 Percentage of stakeholder groups, distinguished by age, with preference for Hamoun use, agriculture, industry and services

Stakeholder group	Number of respondents	Hamoun use	Agriculture	Industry	Services
Young	52	78.8	5.8	11.5	3.8
Middle	48	72.9	6.3	14.6	6.3
Old	46	87.0	2.2	6.5	4.3

It can be concluded that although use of the Hamoun wetland for employment is preferred, there are also quite some people who prefer to move to a different activity, especially to industries. Decision-makers should keep in mind that the Hamoun wetlands may not have sufficient capacity to employ all households, and that industries and services may offer good alternatives.

Over-exploitation

Over-exploitation will affect the availability of goods and services in future years. Penning and Beintema (2006) identify over-exploitation of resources as one of the causes of the current degradation, which started in the late 1980s. They attributed this to two causes: 1) increased population pressure, and 2) the release of exotic fish species. Especially these exotic fish species are believed to have caused wetland degradation, since they feed on the roots of the reeds. To prevent over-exploitation, use should be limited. A rule of thumb could be that not more than 30% of the population of birds and fish is to be caught. For reed, a rule could be that after droughts reeds may be harvested in the second year, to give the plants time to recover. Legislation and enforcement are required to ensure that people will limit the catches and harvests. If such legislation is applied, the relationship between ecosystem condition and use of resources, and hence income and food, will be affected. Currently, such legislation is not available. Over-exploitation therefore mainly affects the relationship between hydrological parameters and ecosystem condition, and will be discussed further in the section dealing with that topic (section 5.5.6).

Approach to quantify indicator values for income and food
As indicator to show changes in income & food, the return period of years with less than 50% of the income in the reference situation is chosen. Changes in income are not expressed in absolute monetary levels, but in percentages, for two reasons: 1) people making use of natural resources and agriculture may have only a small monetary income, because they obtain many products directly from the wetland, field or livestock for their own use, and 2) people may not like to give detailed information about their income. Moreover, it is the difference in well-being between various situations that is of interest in this thesis, and the percentages are highly useful for this purpose.

To calculate the return period of years with less than 50% of the reference income, at first the change in income for each stakeholder group needs to be assessed. This is done as follows:

1. Assessing the availability of resource per resource user as percentage of the availability in the reference situation, which is taken as 100%.
2. Estimating the resulting income as percentage of the reference situation through multiplying the contribution of income from a resource with the resource availability:
    a. condition of reeds is used to calculate the income from reed harvesting and from animal husbandry on pastures;
    b. condition of fish is used to calculate the income from fisheries;
    c. condition of birds is used to calculate the income from bird catching;
    d. condition of agricultural crop production is used to calculate the income from field cultivation (non-fodder crops) and from animal husbandry on farm (fodder crops).
3. For each group the resulting income from activities is summed to calculate the total change in income for each stakeholder group.

5.5.3 Health
Link with ecosystem
Each year between May and August the Sistan area is prone to the notorious '120-days wind', with an average force of 5-6 Beaufort. When the Hamoun wetlands are

dry during this period, these winds are laden with sand and dust. Sandstorms are associated with respiratory, eye and skin complaints. Table 5.12 shows a considerable increase of people suffering from one of these complaints in the situation before and during drought. Sandstorms are also more generally perceived as a nuisance: sand enters the houses and moving sand dunes have even covered houses and parts of villages.

As a service which is not consumed, sandstorm protection can benefit the entire population without the risk of being over-exploited. The sandstorm protection function is largely determined by the presence of vegetation (reed) and water.

Context

Human health is not only affected by the sandstorms. Other factors related to the drought may have led to health complaints, such as the decreased humidity and increased temperature. Also changes in income and food production may have led to a decreased availability of fresh and sufficient food of all nutrient groups. It is not possible to separate the different factors which affect peoples' physical health from each other. Income and food are dealt with as a separate value in this study, additional health impacts are assumed to be related to sandstorms.

Table 5.12 Percentage of stakeholder group suffering from respiratory, eye and skin complaints in the situation before and during drought

	Before drought			During drought		
	Respiratory complaints	Eye complaints	Skin complaints	Respiratory complaints	Eye complaints	Skin complaints
Bird catchers	0	0	0	11	8	4
Fishermen	0	0	0	15	12	6
Reed harvesters	0	0	0	8	15	4
Pastoralists	1	0	0	12	10	5
Field cultivators near Hamoun	0	0	0	16	11	3
Animal farmers	0	0	0	14	16	7
Field cultivators far from Hamoun	0	0	0	9	6	2
Urban people	0	11	0	10	8	11

A second aspect of the context with respect to this well-being value is the exposure to these sandstorms, which could be different for different groups of people. Although it was expected that people living closer to the Hamoun wetlands were possibly more exposed to the sandstorms, the data do not show a higher incidence of complaints among the five groups living close to the Hamouns (bird catchers, fishermen, reed harvesters, pastoralists and field cultivators near the Hamoun) compared to the other three groups. Apparently, the entire area suffers from the sandstorms.

Chapter 5

Approach to quantify indicator values for health

The main link between health and water resources management (apart from income & food) is the exposure to sandstorms. No differences were found in exposure to sandstorms between different stakeholder groups. The exact relationship between human health and the occurrence of sandstorms is not known. Because this relationship is the same for all stakeholder groups, the return period of years with increased sandstorm risk is chosen as indicator. Of the three Hamouns, Hamoun-e-Saberi plays the main role in the protection against sandstorms, because of its location northwest from the populated area. With the prevailing northwest direction of the wind, sand from a dry Hamoun-e-Saberi bed is blown straight into the villages. Therefore, a year is counted as having increased sandstorm risk, when during the period from May through August the inundated area of Hamoun-e-Saberi becomes small. It is assumed that this will happen when the maximum inundated area of Hamoun-e-Saberi during a year is less than 20% of its maximum area.

5.5.4 Perception & experience

Link with ecosystem

All stakeholder groups value the use of the wetland for recreation, which mainly seems to consist of picnicking and enjoying the scenery. The people living close to the wetlands use the area also for traditional and religious ceremonies (see Table 5.13). Which characteristics of the wetland make it suitable for recreation and ceremonies was not investigated. Probably the combination of visual aspects: water, reed and birds, will contribute to the appreciation of the area as perception & experience value.

Table 5.13 Importance of Hamoun wetland for recreation and religion & traditions before and during drought

Stakeholder group	recreation before drought	recreation during drought	religion and traditions before drought	religion and traditions during drought
Bird catchers	100	0	84.6	0
Fishermen	100	2.4	90.2	2.4
Reed harvesters	100	0	96.4	0
Pastoralists	100	0	91.9	0
Field cultivators near Hamoun	94.7	0	84.2	0
Animal farmers	100	0	100	0
Field cultivators far from Hamoun	100	0	53.8	0
Urban people	100	0	32.3	0

Context

The context of this component of well-being would consist of alternative opportunities for recreation or for experiencing religious or other traditions. The Hamoun wetlands are probably the only green and comfortable location in this desert area of Iran. Because these recreation and ceremonial uses are given less priority by the population, the differences in context for the different stakeholder groups are assumed to be small. Changes in ecosystem condition were therefore understood as affecting all stakeholder groups in the same way.

Approach to quantify indicator values for perception and experience

Similar to the relationship between health and sandstorms, the exact relationship between enjoying recreation and condition of the wetland is unknown. However, also here, this relationship is similar for all stakeholders. Decided is to use the ecosystem condition itself as indicator. Of the total ecosystem, it will be the visual components contributing most to the perception and experience value: water, reeds and birds. Of these three components, reeds was chosen as indicator. The presence of reeds implies the presence of water. The presence of birds will contribute to the perception, but also without birds the wetland will contribute to the perception & experience of the population.

5.5.5 Secondary impacts

Link with ecosystem

Secondary impacts are linked with the ecosystem through changes in the other values. An example of such a secondary change is the need for nomads to settle, as a result of the reduced availability of domestic water from the Hamoun wetland (the direct impact). Previously, the nomads lived together with large groups of relatives and shared all their belongings, nowadays the relatives are divided into various households with their own income and expenditures. Public water supply, at the other hand, has prevented people from migration which may have had even more severe secondary impacts through the changes in social structure and loss of family ties. Another example of secondary impacts is the increased dependence and reduced self-esteem, resulting from loss of income and the need for assistance by others or the need to involve in illegal activities.

In case of the Hamoun wetlands secondary impacts can be summarised as follows:

- Social structure:
    - settlement for nomads – related to health value (domestic water);
    - migration to other areas – related to income and food.
- Independence:
    - dependence on aid from government or relatives – related to income and food;
    - dependence on illegal activities – related to income and food.

Context

The relevant aspect of the context for this value is the different extent to which the identified stakeholder groups are affected, and especially how they experience this. Of the identified stakeholder groups, the nomads have experienced the largest changes in their way of living. Data on migration of possible stakeholders to other areas was not collected.

Approach to quantify indicator score

The focus of this case study is on the direct impacts. Assessing secondary impacts and related human well-being values would require additional in-depth studies to understand how the mental well-being and social structures are affected through changes in the human well-being values which are directly affected by the ecosystem changes. Because of this, the secondary impacts are not taken into account in the quantitative assessment in this case study.

5.5.6 Conclusions

This section discussed how each of the well-being values is linked to the Hamoun wetlands ecosystem for the identified stakeholder groups, based on which indicators were identified and the approach for quantification discussed. For income-generation, various aspects of the context are relevant resulting in different relationships for different stakeholder groups. For the other two values, the context showed little differences in the relationship between ecosystem condition and well-being value between the identified stakeholder groups. Because for these values the exact relationship between ecosystem and well-being was hard to assess, but the relationship would be the same for all stakeholder groups, the ecosystem condition itself is selected as indicator. For health, this indicator is the return period of years with increased risk for sandstorms, and for perception & experience the condition of the reed beds.

Table 5.14 Overview of indicators and quantification approach for each well-being value

Well-being value	Indicator	Approach
Income & Food	Change in income (%) with respect to the reference situation	Change in income related to the change in availability of relevant resources (reeds, fish, birds, agricultural crops)
Health	Risk of occurrence of sandstorms (return period)	Return period of years during which not more than 20% of the maximum area of Hamoun-e-Saberi is inundated
Perception & experience	Availability of reeds (%) with respect to the reference situation	Availability of reeds (%) with respect to the reference situation

5.6 Relationship between Hamoun wetlands ecosystem and the Hamoun hydrology

5.6.1 Link with flow regime: hydrological parameters

This section discusses the links between the main ecosystem goods and services (reeds, fish, birds and the prevention of sandstorms) and the hydrological regime of the wetland. Based on analysis of the reference situation (1970s), Penning and Beintema (2006) identified hydrological parameters which they believed are of key importance for the ecosystem condition. Because the ecosystem condition cannot be determined for all possible values of these parameters, thresholds are identified which delineate classes of parameter values. Each possible parameter value, resulting from a water allocation strategy, will be part of one of the classes. A distinction is made between hydrological variation on an inter-annual and an intra-annual scale.

Inter-annual variations

Large floods and droughts have always been part of the Hamoun wetland ecosystem, and have resulted in the sustenance of endemic species. Without frequent floods and droughts, there is a risk that exotic species take over. Besides this, flushing of the area is important to maintain the ecosystem as a freshwater ecosystem. Lakes at the end of an inland basin will inevitably become saline over time. Although the Sistan delta has been referred to as a closed inland delta, it is not the real end of the river system. During large floods the Hamoun wetlands spill into the Shile river, which ends up in the Goud-e-Zereh lake in Afghanistan. This salt lake is the real end of the river system. Without this regular flushing, the Hamoun wetlands will become the end of the system, and hence saline.

The first inter-annual parameter is therefore spilling of the Hamoun wetlands. Penning and Beintema (2006) estimate spilling to have occurred every 8-15 years. Two classes can be defined: 1) a return period of spills smaller than or equal to 15 years and 2) a return period of more than 15 years. This parameter is defined for each of the three Hamouns.

The second inter-annual parameter is a complete drought. This is mainly required to ensure that possibly present exotic species are removed and to force reduction of livestock numbers. In a situation with good management, when enforcement of legislation takes place, occurrence of complete droughts is no longer required. For calculation of complete droughts it is necessary to consider the entire Hamoun area. As threshold is chosen that this entire area has at least during one month an area of less than 10% of the total area. This should happen every few decades. The required return period is estimated at 20 years. Two classes can be defined: 1) droughts take place and 2) droughts do not take place.

Summarising, two inter-annual hydrological parameters can be defined:

Parameter 1: flushing of the Hamoun lakes.
Threshold: once every 15 years each individual Hamoun should have a spill (for Hamoun-e-Puzak and Hamoun-e-Saberi this will mean a spill into the next Hamoun, for Hamoun-e-Hirmand this will mean a spill into the Shile).

Parameter 2: complete drought.
Threshold: once every 20 years, there should be a moment that the total area of all Hamouns is less than 10% of the maximum area.

Intra-annual variations

The inter-annual variation can be considered a prerequisite for the specific ecosystem to develop at all. The intra-annual variation then determines the quality of this ecosystem. The hydrological year in this part of Iran is from October till September (Figure 5.3). Water levels start rising from November or December, while large floods normally arrive between February and April. The wetland reacts with a delay of one to two months. After the large flood has filled up the wetland, the area starts drying up. Depending on the volume of water arriving in spring, the area will still contain a smaller or larger volume when the discharge starts increasing again in fall. The total area in fall and winter will be the result of floods of the previous years as well as of inflows of the new hydrological year.

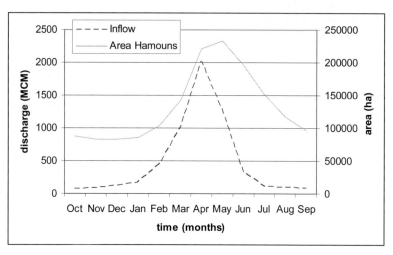

Figure 5.3 Average monthly inflow into Sistan and wetland areas

A distinction of three, partly overlapping, seasons is considered relevant for the wetland ecosystem (Meijer & Van Beek, 2006):

1. October till January: the period when migratory birds visit the area;
2. February till June: the growing/breeding season for birds, fish and vegetation;
3. May till September: the water availability is of less importance for the ecosystem, but a dry bed of the Hamoun-e-Saberi between May and August will increase the risk for sandstorms.

On an intra-annual basis, for the ecosystem it is mainly important that there is sufficient inflow of water during the growing season and that there is sufficient lake area in fall for migratory birds. In addition, it is important that there is some discharge in the rivers to allow upstream migration of fish.

Therefore the following parameters were selected:

Parameter 3: total volume entering the wetland from February to June.
Threshold: the total volume which should be entering the wetland between February and June is put at 2000 MCM. This amount is required for the three Hamouns together.

Parameter 4: inundated area during October-January.
Thresholds: preferably 40% or more of the area should be inundated. This is the result of the size of an average spring flood in the previous hydrological year. When the inundated area is below 40% of the total area, but still above 20% the effects on the ecosystem are less severe, than when the inundated area falls below 20%.

As was discussed in section 5.5.3, the 120-days wind generally blows between May and August from a northeast to southwest direction. The occurrence of sandstorms can be directly related to availability of water in the Hamoun wetland, without considering the ecological consequences. Presence of vegetation will contribute to stabilisation of the soil, even when the plants are dead. However, the central areas of the lakes are without vegetation. When these areas are dry during the period of the 120-days wind, there is an increased risk for sandstorms. Because of the prevailing northwest direction of the wind, the Hamoun-e-Saberi is most important in protecting the Sistan delta from sandstorms. Therefore, as a fifth parameter the inundation extent of the Hamoun-e-Saberi was chosen.

Parameter 5: inundation extent Hamoun-e-Saberi.
Threshold: 20% of the area should be inundated between May and August.

5.6.2 Context

The relationship between hydrological parameters and ecosystem condition was estimated through expert judgement. Also this relationship is influenced by the context. The following aspects of the context were identified:

- water quality;
- over-exploitation;
- mismanagement.

Water quality

Pollution through waste water is not considered a problem for the Hamoun wetland. Most of the drainage water from agriculture and households is isolated from the Hamoun wetlands. The main water quality problem would be the increase in salinity when the area would not be flushed regularly. For that reason flushing floods are considered a constraint. Thus, this aspect of context is internalised in the relationship between hydrological parameters and ecosystem condition.

Over-exploitation and mismanagement

Analysis of satellite images of water volume and reed cover of the Hamoun wetlands over time (See Figure 5.4 and Figure 5.5), revealed that the reed beds decreased before the drought started. Also the placement of the Hamoun wetlands on the Montreux-list, a list of threatened Ramsar-sites, in the 1990s indicates that degradation started before 1997, which is the year the drought started. Penning and Beintema (2006) attribute this to over-exploitation and mismanagement of the Hamouns.

The fact that the ecosystem is valued for its goods implies that the ecosystem is exploited. Harvesting products from an ecosystem affects its condition. When this is done in a sustainable way, the ecosystem recovers and goods and services remain available. With the large growth of the population, from 174,000 up to 404,000, pressures on the wetland most likely increased, with over-exploitation as a result. Under over-exploitation, the relationship between hydrological parameters and presence of reeds, birds, and fish will change. The effect of over-exploitation will become apparent in the subsequent year. Catches of fish and birds should be restricted to a percentage of the population, while for reeds use should mainly be restricted during the first year after a period of degradation.

Chapter 5

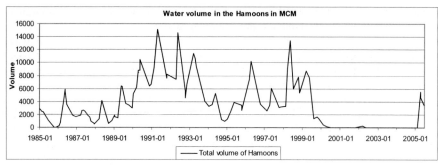

Figure 5.4 Water volumes in the Hamoun wetlands, 1985-2005 (Source: Vekerdy & Dost, 2005)

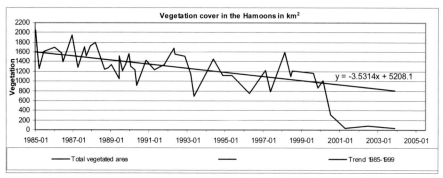

Figure 5.5 Vegetation cover in the Hamoun wetlands, 1985-2005 (Source: Vekerdy & Dost, 2005)

Related to over-exploitation is the introduction of exotic fish species, the grass carp (*Ctenopharyngodon idella*), in the 1980s. These fish feed on the roots of the reeds. As a result, reed beds were destroyed, with subsequent reduction in the bird population.

Currently, there is no sign that over-exploitation is prevented and that exotic species are no longer released, although the negative impact of the grass carp is gaining awareness. This is the reason for including the occurrence of complete droughts as a hydrological parameter. If good management will be implemented, this hydrological parameter is no longer of relevance for sustenance of the Hamoun ecosystem.

5.6.3 Quantifying the Hamoun ecosystem and hydrology relationship

In this section the link between the Hamoun wetlands ecosystem and the identified hydrological parameters is quantified. Because the ecosystem goods and services of reeds, fish and birds all depend on various hydrological parameters, as well as on changes in the goods and services themselves, there is no straightforward link between a change in a single hydrological parameter and these goods and services. Instead, the changes in the various parameters should be considered in total as a change in the hydrological situation of the Hamouns. This does not apply to the service of sandstorm protection, which was defined as having a straightforward relationship with inundation of the Hamoun-e-Saberi. In the discussion in this section, the fifth parameter of inundation of the Hamoun-e-Saberi between May and August is therefore not considered. Combining the remaining four parameters and thresholds leads to 24 theoretically possible hydrological situations (Table 5.15). Of these, nine combinations were considered not possible to occur.

Ecological impacts are always uncertain, and especially with the limited data available in this case. Similar to the approach of DRIFT, changes are expressed as a range of possible change. For each combination of parameter values, the resulting conditions of reeds, fish and birds are estimated as percentages of the condition in the reference situation. The estimations are made through expert judgement based on the analysis of the ecosystem by Penning and Beintema (2006). Table 5.15 presents the estimated lower and upper boundary of ecosystem condition for each combination of parameter values. This table with estimated ecosystem impacts will be used to assess ecosystem impacts of changes in the Hamoun hydrology

Table 5.15 Estimated effects on reeds, fish and birds for combinations of parameter classes

Inter-annual		Intra-annual		Reeds		Fish		Birds	
Spill	Drought	Flood spring	Area fall	Lower bnd*	Upper bnd*	Lower bnd*	Upper bnd*	Lower bnd*	Upper bnd*
T ≤ 15 y	T ≤ 20 y	≥ 2000 MCM							
yes	yes	yes	>40%	100	100	100	100	100	100
			20-40%	90	100	90	100	90	100
			<20%	90	100	20	90	50	90
		no	>40%	**	**	**	**	**	**
			20-40%	50	100	30	80	90	100
			<20%	30	70	20	70	50	90
	no	yes	>40%	10	50	0	50	90	100
			20-40%	10	50	0	50	80	100
			<20%	10	50	0	50	40	80
		no	>40%	**	**	**	**	**	**
			20-40%	10	50	0	50	70	90
			<20%	0	20	0	20	30	70
no	yes	yes	>40%	60	80	0	0	90	100
			20-40%	40	60	0	0	80	100
			<20%	20	40	0	0	40	80
		no	>40%	**	**	**	**	**	**
			20-40%	20	40	0	0	70	90
			<20%	0	20	0	0	30	70
	no	yes	>40%	**	**	**	**	**	**
			20-40%	**	**	**	**	**	**
			<20%	**	**	**	**	**	**
		no	>40%	**	**	**	**	**	**
			20-40%	**	**	**	**	**	**
			<20%	**	**	**	**	**	**

* boundaries (bnd) are expressed in percentage of the "healthy" 1975 reference situation
** assumed to be impossible combinations of parameters

5.7 Relationship between the Hamoun wetland hydrology and the upstream flow regime

5.7.1 Link with upstream flow regime

A relationship between upstream and wetland flow regime is required to estimate the impacts of changed upstream flow regimes, resulting from water resources management strategies, on the hydrology of the Hamoun wetlands. Such strategies can impact the flow regime at various locations. Therefore, in this case study, the flow regime is relevant at various locations upstream of the Hamoun wetlands. Because not much information is available on the basin in Afghanistan, the focus on linking upstream and local flow regime in on the basin in Iran. With the generated information, Iran can obtain insight both in how their own water resources management affects the Hamoun wetlands and in how much water they require as inflow from Afghanistan. The Iranian part of the river basin is modelled as an application of RIBASIM.

5.7.2 Context

All factors which affect the relationship between the upstream flow regime and wetland flow regime are included in the RIBASIM application. Therefore, there is no need to separately consider the context. The RIBASIM application will be discussed briefly in the following section. The application is discussed in more detail in Meijer and Van Beek (2006).

5.7.3 Quantifying the wetland hydrology and upstream flow regime relationship

RIBASIM schematisation

RIBASIM is a water balance model with which a water resources system can be modelled through a network of nodes and links. All nodes represent either inflow, water use, or a control structure such as a weir, a bifurcation or a reservoir. The links mainly transport the water between the nodes in an upstream-downstream direction. Priorities can be added to overrule the upstream-downstream water allocation.

With such a water balance model the impact of water resources management strategies can be assessed for a long range of (historical) inflow data. The schematisation represents a specific water resources management situation, or strategy, with specific water demands and operation rules. With the water balance model, it can be tested whether with this strategy all demands are fulfilled both in years with high and with low inflow. This requires a long series of inflow data, which should represent all natural variations in high and low flow.

For the Sistan application inflow series were available for 55 years. These series are a combination of available measurement data for 55 years for the Hirmand river at Hirmand Fork, and of generated series for Farah and Khash based on statistical analysis of shorter series of measurement data. Inflow from the Adraskan and Khaspas are assumed to be of minor importance compared to the inflow from Hirmand, Farah and Khash, and were not included. With these inflow series expected water resources developments in Afghanistan were simulated in a separate RIBASIM application to derive inflow series for a future situation. This RIBASIM application for Afghanistan is discussed in Kwadijk and Diermanse (2006). As simulation timestep for the RIBASIM applications one month was chosen. This was considered a suitable timestep to estimate the values for the identified hydrological parameters, which all take place at the scale of seasons to years.

Representation of Hamoun wetland hydrology in Ribasim simulations

The RIBASIM application for Sistan was calibrated for both the reference situation (1975) and the present situation (2005). Two types of data were used to check the simulation results and calibrate the model: 1) return period of spills from Hamoun-e-Hirmand into the Shile river and 2) observed inundated areas from satellite images. More information on these data can be found in Penning and Beintema (2006). The calibration itself is discussed in Meijer and Van Beek (2006). Here only the results of the calibration are presented for the situation with water use and management of 1975 (further referred to as reference case) and of 2005 (further referred to as present case). Please note that for both simulations the same set of inflow series of 55 years was used.

The only data available about the return period of spilling of the Hamouns is the estimation that these spills used to occur from the Hamoun-e-Hirmand into the Shile river every 8-15 years. To maintain the low salinity levels of the Hamoun-e-Puzak and Hamoun-e-Saberi, these wetlands require spilling as well. Table 5.16 shows that both in the reference case and in the present case spilling of all Hamouns occurs with a return period of less than 15 years. The return period of spilling of the Hamoun-e-Hirmand has increased (less frequent spilling) in the present case compared to the reference case. This can be explained by the fact that the developments since the 1970s (construction of Chahnimeh reservoirs and increase of irrigated area) have mainly taken place along the Sistan river. The Hamoun-e-Hirmand is mainly fed by the Sistan river, while the other two Hamouns depend more on the inflows from the Common Parian, the Farah and the Khash.

Table 5.16 Return period of spills for each Hamoun

	Data	Reference case	Present case
Hamoun-e-Puzak		6	6
Hamoun-e-Saberi		11	11
Hamoun-e-Hirmand	8-15	8	14

Figure 5.6, Figure 5.7 and Figure 5.8 present the calibration results for the inundated area of the three Hamouns in the reference and the present case. The Hamoun-e-Puzak and Hamoun-e-Saberi have not been affected much by the increase in water use between 1975 and 2005: the lines of the reference and the present situation lie on top of each other. However, the inundated area of Hamoun-e-Hirmand decreased between 1975 and 2005. This can again be explained by the developments that took place along the Sistan river between 1975 and 2005.

The measured data show a decline in inundated area at the end of the 1990s. The simulation results for the Hamoun-e-Hirmand also shows this decrease, but the reduction is not found in the simulation results of the Hamoun-e-Puzak and Hamoun-e-Saberi. The reason for this difference between the three Hamouns can be explained by the fact that Hamoun-e-Hirmand receives most of its water from the Hirmand river (through the Sistan river), while the other two Hamouns depend more on the other inflowing rivers. Since the inflow series for the Farah and the Khash are generated based on statistics, they do not include this drought in the 1990s. This means that the simulation results for the Hamoun-e-Puzak and Hamoun-e-Saberi, and to a lesser extent also for the Hamoun-e-Hirmand, cannot reproduce the exact past hydrological regime, but they do show the natural variations, and are therefore considered suitable for analysis of water resources management strategies.

In the Hamoun-e-Hirmand, the decline in inundated area at the end of the 1990s is observed in the simulation results of both the reference and the present case. The fact that also in the reference situation the low inflows at the end of the 1990s would lead to a reduced Hamoun area indicates that the reduced Hamoun inundation is the result of low inflows from Afghanistan and not from water use within the Iranian part of the basin. These low inflows, in turn, can be caused either by a climatological drought or by a large increase of withdrawal in Afghanistan. Because of the present instable political situation in Afghanistan, it is not very likely that large scale infrastructural developments have taken place in recent years. A climatological drought is therefore more likely.

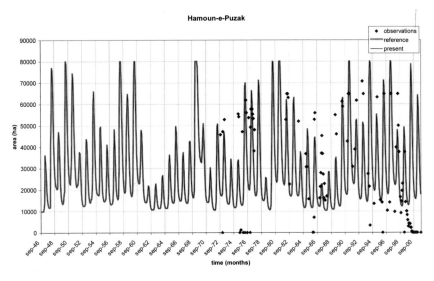

Figure 5.6 RIBASIM calibration results for the water coverage of Hamoun-e-Puzak for the reference and the present case

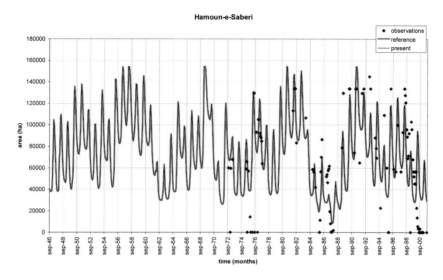

Figure 5.7 RIBASIM calibration results for the water coverage of Hamoun-e-Saberi for the reference and the present case

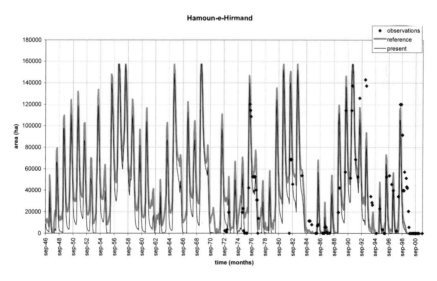

Figure 5.8 RIBASIM calibration results for the water coverage of Hamoun-e-Hirmand for the reference and the present case

This calibrated RIBASIM application is used to translate water resources management strategies into hydrological parameter values for the Hamouns. For the calibration of the RIBASIM application only limited data were available. The simulation results can therefore not be considered to describe exactly what will happen as a result of a particular strategy, but will provide good insights in the different impacts of the Hamoun wetlands as a result of alternative strategies.

5.8 Combining all: Impacts of changed Hamoun hydrology on the well-being of the identified stakeholder groups

The previous sections discussed all links of the conceptual model. The purpose of assessing these links is to be able to quantify the impacts of changed flow regimes, resulting from water resources management strategies, on human well-being. Following the way of thinking in the 'scenario-based' environmental flow assessment methods, the desired condition of the ecosystem, and hence the environmental flow regime, follows from analysis of all impacts of changes in the flow regime. The selection of the best strategy, which defines the desired ecosystem condition, follows from political decision-making. This chapter provides information that could contribute to such a decision-making process.

The approach used to assess the impacts of human well-being is the following. Alternative water resources management strategies were simulated with RIBASIM to find the resulting changes in the Hamoun hydrology. Post-processing of the RIBASIM output was required to assess the values for the four hydrological parameters of Table 5.15. The combination of the values for the inter-annual parameters forms the first step to determine the ecosystem condition from Table 5.15. Six possible situations will remain based on the intra-annual parameters. For the intra-annual parameters there are two options: 1) calculate the average volume/area over the simulated years, and 2) count the number of years during which a certain combination of parameters takes place. The average condition is useful to have an impression of the average ecosystem condition after a long range of years. However, the average does not show that there may be many years during which the ecosystem provides only limited goods and services. Therefore, to assess the socio-economic impacts, counting the years during which a certain situation takes place is preferred. This way insight is obtained in how many years the ecosystem goods and services are insufficient to meet the needs of the population.

The following sections discuss at first some possible water resources management strategies which will affect the flow regime, and subsequently the impacts these strategies will have on the hydrology of the Hamoun wetlands, the availability of resources from both the wetlands and irrigated agriculture and finally what this implies for human well-being.

5.8.1 Water resources management strategies which affect the flow regime

To improve the water resources management system of Sistan, various measures can be taken. Measures should preferably be defined in dialogue with the responsible authorities and other stakeholders. At the time of writing of this case study, the IWRM study of the Sistan closed inland delta was still ongoing, and no measures were discussed yet. Therefore, the analysis discussed in this section is meant to provide a first insight in how possible measures affect the wetland hydrology, the ecosystem condition and subsequently the well-being of the identified stakeholder groups.

The two main water users in the Sistan delta are irrigated agriculture and the Hamoun wetlands. The selected strategies therefore focus on alternative distributions of water between these two demands. Two strategies were identified:

1. Increase of the agricultural area to 245,000 ha: extension of the irrigated area is an existing plan of the Ministry of Agriculture to improve the socio-economic situation in the area.
2. Decrease of the agricultural area to 21,000 ha: to protect the ecosystem decrease of the irrigated area is suggested.

Both strategies assumed that good ecosystem management is applied, which means that the occurrences of complete droughts are no longer a criterion in assessing ecosystem condition.

5.8.2 Impacts of water resources management strategies

The results of strategies are presented in Table 5.17, together with the results for the reference and the base case. The focus of the analysis is on the impact on human well-being. Because for virtually all stakeholders their income depends both on Hamoun goods and services and on agricultural production, the table displays the availability of ecosystem and agricultural resources. Agricultural production was not part of the case study itself. Agricultural water demands and yields were investigated in the IWRM study and follow directly from the RIBASIM simulations (See Meijer and Van Beek (2006) for information on agricultural data and analysis). These results were used to quantify changes in income for the various stakeholder groups. The hydrological results in the table are the resulting values for the hydrological parameters which were identified as important determinants of the ecosystem condition.

Hydrological impacts Hamoun ecosystem condition

There is not much difference between the reference case and the case for the present situation. Flushing of the Hamoun-e-Hirmand takes place less frequent in the present case and the inundated area in fall is a little smaller. Increase of the irrigated area results, as expected, in a higher frequency of droughts and a lower frequency of Hamoun flushing, the volume of water reaching the Hamouns in spring as well as the area still inundated in fall are both smaller than in the present situation. When the agricultural area is decreased the opposite occurs.

Resources related to ecosystem condition

The resulting values for the hydrological parameters show variations between the strategies, but because of the use of thresholds, the variations in ecosystem condition are very small. In most cases under the ecosystem condition is fairly good. When the irrigated area is increased, the reduction in flushing of the Hamoun-e-Hirmand results in a reduced ecosystem condition in this part, but is still calculated at 60-90% of the healthy condition.

Chapter 5

Table 5.17 Impacts of water resources management strategies on hydrology, resources and human well-being

Strategies		Unit	Reference case 1975	Base case 2005	1 increase irrigated area	2 decrease irrigated area
Hydrology	**Hydrological parameters**					
	• Complete drought (inundated area <10% of maximum inundated area)	T*	10	10	7	50
	• Average maximum fall area (between October and January)	ha	102,000	90000	79000	119000
	• Average volume entering Hamouns in spring (Feb-June)	MCM	4,500	4200	3800	4800
	• Flushing of Hamouns					
	− Puzak	T	6.3	6.3	6.3	6.3
	− Saberi	T	10.0	10.0	12.5	7.1
	− Hirmand	T	7.1	12.5	16.7	7.1
Resources	**Ecosytem condition**					
	• Reed	% of ref	90-100	90-100	73-87	90-100
	• Fish	% of ref	90-100	90-100	60-67	90-100
	• Bird	% of ref	90-100	90-100	87-100	90-100
	Agriculture					
	• Production fodder	ton/y	67,000	76,000	111,000	18,000
	• Production non-fodder	ton/y	324,000	325,000	459,000	93,000
Human well-being	Income & food (return period of years with <50% of reference income)					
	• Bird catchers	T	n/a**	5.44	1.69	9.80
	• Fishermen	T	n/a	5.44	1.00	8.17
	• Reed harvesters	T	n/a	5.44	1.09	9.80
	• Pastoralists	T	n/a	5.44	1.69	4.90
	• Field cultivators near Hamoun	T	n/a	2.04	3.77	1.00
	• Animal farmers	T	n/a	1.20	n/a	1.00
	• Field cultivators far from Hamoun	T	n/a	1.20	n/a	1.00
	• Urban people	T	n/a	n/a	n/a	n/a
	Physical health (years that Saberi <20% of max area)					
	• All stakeholders	T	12.5	12.5	7.1	50
	Perception & experience (average condition of reeds)					
	• All stakeholders	% of ref	90-100	90-100	73-87	90-100

Resources related to irrigated agriculture

In the current situation, irrigation water demands are the highest priority after public water supply. Agricultural production is therefore mainly related to the area irrigated. In the reference case this is 85,000 ha, which increased till 120,000 ha in the present case. Further increase upto 245,000 ha gives a higher production, while, logically, the production of both fodder and non-fodder crops is much lower when the irrigated area is decreased. It is noted that doubling the irrigated area does not lead to a doubling of the crop production. This indicates that under this strategy serious water shortages can be expected to take place.

Human well-being

Although the resulting ecosystem condition of the various cases do not show much variation, the resulting well-being impacts do show variations between the strategies. In the reference case, none of the stakeholder groups faces years with less than 50% of the reference income. Although in the present case and in the case with a decreased irrigated area the ecosystem condition is the same as in the reference case, the income of the stakeholder groups is lower as a result of the lower availability of resources per household as a result of population growth. When the irrigated area is decreased, the groups of bird catchers, fishermen and reed harvesters face the least years (largest return period) with less than 50% of the reference income. For the animal farmers and field cultivators far from the Hamoun wetlands this is the case when the irrigated area is increased. Under all strategies certain stakeholders, especially the Hamoun-using groups, will face years with reduced income. Therefore, a discussion with the stakeholders about compensation or mitigation is required.

5.9 Discussion & Conclusions

Conclusions from this case study can be drawn at three levels, following the structure used in Chapter 4:

1. on the impacts of water resources management on Hamoun wetland-related human well-being;
2. on the suitability of the methods used for the application of the conceptual model;
3. on the lessons learned with respect to the usefulness of the conceptual model and approach for the quantification of human well-being impacts in Integrated Water Resources Management.

5.9.1 Impacts of water resources management on Hamoun wetland-related human well-being

With regard to their dependence on the Hamoun wetlands, the population of the Sistan delta can be split into three groups. Of the total of approximately 71,000 households, around 15,000 households depend for more than 70% of their income on the Hamoun wetlands, around 30,000 for more than 70% of their income on irrigated agriculture, while the income of the remaining 26,000 urban households is largely not related to water.

The main goods and services important for the population are the availability of fish, birds and reeds to support income and food production, as well as the prevention of sandstorms and the regulation of the local climate. Five important hydrological parameters were identified to sustain these goods and services. For the sustenance of fish, reeds and birds these are wetland spills, regular droughts, a minimum inundated area in fall, and a minimum volume of water inflow in spring. It was considered important to have a minimum area of Hamoun-e-Saberi inundated from May till August to prevent sandstorms.

Water resources management strategies can affect the well-being of the identified stakeholder groups through their impacts on the wetland hydrological regime, subsequent changes in ecosystem, and resulting availability of ecosystem goods and services. For all changes in ecosystem condition an approach to quantify the change in well-being for the identified stakeholders was discussed. With these links the impacts on the stakeholder groups and the equity in impact distribution was assessed.

The main water demands come from irrigated agriculture and the Hamoun wetlands. Both water use sectors are important for the economy of the region and the well-being of the stakeholder groups. A balance needs to be found in the distribution of water over these two demands. The strategies analysed focussed on investigating this balance. It was found that in the present situation the distribution of impacts is most equitable. Increasing the irrigated area will mainly benefit the farmers, while the Hamoun users will mainly benefit from a decrease of the irrigated area. It should be noted that these are the results of the current models, which, due to a limited availability of data, contain some uncertainties. Further investigation and analysis are required before decisions on water resources management can be made.

Even with the limited data, it was found that over-exploitation was most probably an important factor in the degradation of the ecosystem. Simply supplying water to the wetland may not result in the desired ecosystem condition. Prevention of over-exploitation is an important task for the local authorities. To understand the limits of ecosystem use, monitoring of the ecosystem condition under various hydrological and use situations is required.

What the environmental flow requirement for the Hamoun wetland is, depends on how the value of a certain Hamoun condition weighs out against the benefits of other water use. From the viewpoint of the stakeholder groups who depend on the Hamoun wetland for their income, the preferred strategy would be to decrease the irrigated area, which means a more natural flow regime for the Hamouns. On the other hand, the field cultivators and animal farmers will prefer an increase of the irrigated area. The present situation case seems to spread income effects most equitable over the different stakeholder groups. The results presented here are just a few examples of how the situation will change with possible measures. The choice of the preferred strategy lies with the decision-making authorities.

These local authorities are assumed to represent the local population. However, certain groups are better represented than others: the Ministry of Jehad and Agriculture mainly supports the people involved in irrigated agriculture and animal husbandry. Fishermen are to some extent represented by the Fisheries bureau of the Ministry of Jehad-Agriculture, and pastoralists by the Ministry's Bureau of Nomads. The group of bird catchers and reed harvesters are however not represented by a formal authority.

5.9.2 Suitability of the methods used for the application of the conceptual model

The conceptual model was applied in this case study through the same five steps as in the case study on the Surma-Kushiyara rivers of Chapter 4. Data on the link between human well-being and wetland hydrology and ecosystem condition were collected in two stages. In the first stage, during the group discussions, an understanding was obtained of the important well-being values related to the ecosystem, the goods and services, the hydrological parameters as well as on what groups of stakeholders to distinguish. Additional data on the main links and the relevant context were collected through the questionnaire survey and analysed for each stakeholder group. Together with the assessment of links between ecosystem condition, wetland hydrology and water use upstream as part of the comprehensive IWRM study, the well-being data collection allowed for a quantified assessment of well-being impacts resulting from water resources management strategies. The combination of a qualitative and a quantitative assessment of the well-being – ecosystem links can be concluded to be very suitable for application of the conceptual model.

In future studies it remains of utmost importance to have a thorough and timely cooperation between the different disciplines in an IWRM study. The group discussions and interviews can greatly benefit from information on ecosystem condition and hydrological situations of the past. For example, in this case study it was realised in an advanced state of the project that what was understood as a dry year during the group discussions -completely dry during the entire year-, was hardly compatible with the results of the water balance modelling in which the Hamouns were completely dry only during part of the year. In this study this was dealt with by adjusting the threshold for the drought parameter to ensure that the reference and the present case simulation indeed represented the reference and the present situations. This was allowed because the analysis focused on differences in impacts between strategies and not on exact results. Yet, timely identification of possible inconsistencies can trigger additional data collection and reduce uncertainties in the outcomes of the analysis. Not only can the well-being study benefit from ecological and hydrological analysis, the ecological and hydrological assessment can also benefit from the findings from the group discussions and the interviews, for example through the description of past events by the local population.

5.9.3 Lessons learned with respect to the usefulness of the conceptual model and the approach for the quantification of human well-being impacts in IWRM

The conceptual model with the five step approach proved useful in this case study to structure the assessment. Considering the context was useful to understand the relationships between the wetland ecosystem and the well-being of the identified stakeholder groups. Because the changes in the ecosystem and related well-being had taken place already in the Hamoun wetlands, for certain indicators the relationship could be assessed directly through studying the well-being for different stages of wetland condition and for distinct hydrological situations. Therefore, unlike in the case study on the Surma-Kushiyara rivers, the context was not always required to predict how different stakeholders would be affected, but instead to understand why different groups were affected differently and what measures could be taken to improve their situation.

To assess the relationship between ecosystem condition and wetland hydrology, use was made of the principles of the DRIFT approach. This application has shown that DRIFT fits very well in the system analysis approach of IWRM, and that a well-being assessment can be linked to these approaches. There are two complications in the coupling of 1) the relationship between ecosystem condition and hydrological parameters with 2) the relationship between ecosystem goods and services and human well-being. The first complication is that from an ecological point of view, the ecosystem condition is evaluated as a holistic entity, whereas from a human well-being point of view it is the *amount* of goods produced and the *degree* to which services are provided that is important. Time series of hydrological situations, ecosystem conditions and availability of goods and services are required to establish accurate relationships. The second complication is that a healthy ecosystem can exhibit strong variations in availability of goods and services from year to year, as a result of variations in hydrological regime, especially in (sub-) tropical climates. For the ecosystem itself, these variations are part of a healthy condition; without the variations the ecosystem would be a different ecosystem. However, for the people depending on this ecosystem, the temporarily reduced availability of goods and services may be problematic. In this case study this was accounted for by counting the number of years with reduced availability, instead of considering the average condition of the ecosystem over a long period of time.

The proposed approach should be considered a first overall quantitative assessment of how different stakeholder groups are possibly affected by changes in the flow regime and in the river ecosystem. Because of the broadness of this approach and the focus on a topic about which often very few data are available, this assessment cannot go into much depth. In this first assessment, the main stakeholder groups and ecosystem goods and services that will be affected can be identified. Based on this, it can be decided whether additional in-depth studies by certain specialists, such as human health or nutrition specialists, are required.

The scorecards of the analysis of the impacts of alternative water resources management strategies (Table 5.17) showed large differences between the cases with respect to human well-being impacts, while the resulting ecosystem condition in each of the cases did not show that many differences. This indicates that assessing human well-being impacts provides useful additional information which can enhance informed and equitable management.

The main conclusion from this case study is therefore that an assessment of human well-being impacts is useful and required to increase social equity in IWRM decision-making. The conceptual model and stepwise approach provide a framework for a first structured and quantitative assessment of these impacts.

6 Practical application of the conceptual model: lessons from the case studies

6.1 Introduction

The main objective of this thesis is to develop an approach to assess human well-being values related to environmental flows for practical application in IWRM systems analysis. In the case studies, the conceptual model proved to be a useful tool to facilitate the assessment. The two case studies were, however, very different. In the Surma-Kushiyara river system in Bangladesh, interventions had not yet taken place. Human well-being was investigated to understand the link with the current river ecosystem and how this may be affected when flow regimes would change, in order to understand to what extent the current ecosystem needed to be conserved. In the Hamoun wetlands in Iran, large changes had taken place already. The link between human well-being and the situation before and during the drought was investigated to understand how restoration measures could improve human well-being. This difference in situation of before (further referred to as 'conservation' situation) or after (further referred to as 'restoration' situation) interventions in the river flow regime has led to differences in the methods used for data collection and analysis. It proved that a single prescription of how to derive relationships between flows, ecosystems and human well-being cannot easily be given, and that the conceptual model alone cannot guarantee a good assessment of these relationships.

To apply the conceptual model, both case studies followed the five steps described in Chapter 3:

1. identifying stakeholder groups;
2. assessing the relationship between human well-being and the river ecosystem;
3. assessing the relationship between the river ecosystem and the local flow regime;
4. assessing the relationship between the local flow regime and the upstream flow regime; and
5. estimating impacts of water resources management measures on the well-being of the identified stakeholder groups.

The following sections discuss for each of these steps, the similarities and differences between the practical applications in the two very different cases, and derives general recommendations for future assessments of human well-being in the fields of environmental flows and IWRM.

6.2 Step 1: identifying stakeholder groups

6.2.1 Description and envisaged result

The purpose of this step is to identify all people whose well-being is possibly affected by changes in the river ecosystem as a result of changed flow regimes, and to distinguish different groups. The result of this step is an overview of possibly affected groups of people and their numbers, with an expectation about type and importance of their link with the river ecosystem. Based on this, a decision will be made on which groups of people will be included in the assessment, and which people not. During the activities of step two, additional stakeholder groups may be identified, and the initial division in groups which are likely to experience different impacts can be further refined.

This step consists of two activities:

1. inventory of potential stakeholders and assumed interest;
2. selection of stakeholder groups for detailed assessment.

6.2.2 Inventory of potential stakeholders and assumed interest

Various sources of information can and should be used to identify potential stakeholders: literature, national or regional censuses, interviews with local authorities and non-governmental organisations (NGOs), information on hydrological boundaries and information on ecosystem goods and services. Some remarks can be made about the last three.

Interviews with local authorities and NGOs

Local authorities and NGOs generally represent distinct groups of the population. The authorities and NGOs can provide information on the importance of the river ecosystem for the people they represent, other important issues for these people, the number of people and where to find them. Part of the steps in an IWRM analysis is actor analysis: an inventory of all stakeholders and their objectives, perceptions and resources. Identifying stakeholders of river ecosystem use through interviews with local authorities can be combined with this step in an IWRM analysis. Unfortunately, not all groups are properly represented by authorities or NGOs. For example in Sistan, authorities were present for fishermen, farmers and nomads, but not for bird catchers or reed harvesters. Care should be taken that all stakeholders are included in the analysis.

Hydrological boundaries

In the Sistan area in Iran, the Hamoun wetlands were the only large water bodies in an otherwise dry area. Human communities were concentrated on the Sistan plain and around the Hamoun wetlands. Because of the low local precipitation, dependence on the river and the wetland by these people was clear. However, in the wet north-east of Bangladesh it is much harder to identify what water bodies and water use are related to the inflowing rivers. Hydrological or hydraulic analysis is required to understand what should be regarded as the river ecosystem.

Ecosystem goods and services

A first overview of potential goods and services will be of use to identify stakeholder groups. Ecological expertise is required to understand which goods and services should be considered as originating from the river and wetland ecosystem and which are linked to terrestrial ecosystems. Specialist expertise is also required to identify the less obvious goods and services, and the goods and services that may be relevant for groups of people living further away from the ecosystem. For example, in the case study in Iran, the data of the ecologists showed the enormously far-reaching impacts of the sandstorms originating from a dried-out wetland.

6.2.3 Selection of stakeholder groups for detailed assessment

Based on the inventory, a selection probably needs to be made of which groups will be included in a more detailed assessment. Depending on the scope of the project, some groups may not be included in a more detailed assessment, for example because they are not under the responsibility of the decision-making authority or because the ecosystem can be assumed to be only of minor importance to their well-being. For example, the Hamoun users in Afghanistan were not included in the analysis in Hamoun wetlands case study (Chapter 5), because they were not under the responsibility of the decision-making authorities in Iran.

6.3 Step 2: assessing the relationship human well-being and the river ecosystem

6.3.1 Description and envisaged result

In this step, the link between ecosystem condition and human well-being is assessed. The stakeholder groups to be included follow from step 1, but during the more detailed assessment in this step additional stakeholders or additional sub-groups of stakeholders may be identified. The result of this step can consist of either relationships in the form of knowledge rules, or of tables with for different ecological or hydrological situations the various well-being values for all the different stakeholder groups. A combination of knowledge rules and tables is possible as well.

If sufficient data is available, this part of the assessment can be carried out as a desk study. However, data on population and use of ecosystem goods and services are generally not readily available, and therefore fieldwork will often be inevitable. Because the topic is complex, it is advised that data collection is split and carried out in several sessions. Proposed is a qualitative assessment, to identify indicators for the human well-being values, followed by a second assessment to quantify the links between ecosystem condition and the indicators.

The following activities are proposed for this step:
1. fieldwork 1: obtaining a qualitative understanding of ecosystem – well-being relationships;
2. identification of indicators and approach for quantification;
3. fieldwork 2: collecting information for quantification of relationships;
4. assessment of relationships;
5. description of possible secondary impacts.

6.3.2 Fieldwork 1: obtaining a qualitative understanding of ecosystem – well-being relationships

Group discussions or other participatory approaches are suggested for the qualitative assessment, because of the complexity of the topic and because the focus at this stage is not yet on individual values. Good introductions and sufficient time with the population is important to gain trust. The main purpose of the qualitative assessment is to obtain a general understanding of the importance of the various components of the flow regime and of ecosystem goods and service for people.

It is important to spend sufficient time to find out how the people perceive the different hydrological situations. What is the normal cycle of high and low flows, and how long do these phenomena last? What is understood by flood and by drought? Can different types of floods and droughts be distinguished? What happens when there is a deviation from the normal, or average, situation with regard to the ecosystem and the use of the ecosystem? How do people respond to this?

In the restoration case of the Hamoun wetlands in Iran, where people had experienced a drought of seven years, it was easy for people to identify different situations and describe the impacts. In contrast to this situation, the inhabitants of the Surma-Kushiyara floodplain in Bangladesh had not experienced yet any negative impacts of changed flow regimes and had difficulty identifying different flow regime characteristics and a relationship with their well-being. Support from ecological specialists is required especially in the conservation cases to understand whether the goods and services people make use of are indeed related to the river ecosystem.

6.3.3 Identification of indicators and approach for quantification

With the results of the qualitative assessment, the qualitative links with the river ecosystem are understood and indicators for each well-being value can be identified. In the next step, these links and indicators will be quantified.

Identifying operational indicators also involves the choice of assessing the relationships as a value relative to the value in the current or a reference situation, or as an absolute value. Especially for the category income and food, it will be difficult to assess absolute values. People may not like to reveal their total amount of income, and in case of a (partly) subsistence economy, they may not even know their actual income. In such a case it is easier to estimate the change in their income as a result of a change in ecosystem condition, than to assess actual income in different situations.

An important difference between assessing environmental flows for nature and for people is that whereas extreme situations of droughts and floods are important to maintain the natural ecosystem condition over a long period of time, for people the availability of goods and services in each single year is important. During years of scarce goods or unhealthy situations, to which people need to respond, the long-term average is of less importance. Therefore, it is suggested to define indicators for human well-being not as an average over a long period but as the frequency during which a certain availability or shortage of particular goods and services takes place.

Because the quantification requires input from the results of the ecological and hydrological analyses, it is important to communicate what type of input is required and can be provided and to adjust the relationships that need to be quantified accordingly. The ecological analysis may focus on the condition of the ecosystem, whereas for the assessment of human well-being impacts, actual amounts of fish, birds and other goods are desired. In this thesis this problem was solved by assuming that a change in ecosystem condition meant a similar change in ecosystem use, for example in income, without looking at the number of fish available and fish caught. The relationship between ecosystem condition and availability of goods and services is not always linear, however. When the ecosystem provided goods in abundance in the natural situation, a certain amount of ecosystem degradation can occur before ecosystem use by people is actually affected. On the other hand, in cases where the ecosystem is already under pressure, which was assumed to be the case in the Hamoun wetlands because of the fact that degradation was signalled before drought, each degradation of the ecosystem condition will lead to a decrease in amounts available for human use.

An alternative solution when the ecological analysis cannot provide the required data is to link human well-being values directly to a hydrological characteristic such as flood extent or magnitude of spring floods. The population may not recognize these characteristics but may use their own description of hydrological situations such as normal or dry. The social scientists and hydrologists should find out which hydrological characteristics match the description by the population.

When there is a straightforward relationship between ecosystem condition and well-being, and this relationship is the same for all identified stakeholder groups, the ecosystem condition itself can be used as an indicator.

6.3.4 Fieldwork 2: collecting information for quantification of relationships

When the indicators and the relationships that need to be quantified have been identified, additional data need to be collected for the actual quantification. This activity focuses on the importance of various ecosystem goods and services or hydrological situations for the different stakeholder sub-groups.

In the relationships between ecosystem condition and human well-being, the context is important to identify or understand different relationships for different stakeholder groups. Also, differences in context may lead to a further distinction of stakeholder groups, for example in fishermen with and without full access. This further distinction will follow from the qualitative assessment. For the quantitative assessment all possible distinctions in groups should be envisaged in order to ensure that the collected data can be analysed separately for each stakeholder group.

For the consideration of the context, there is an important difference between restoration and conservation cases. In case of a conservation study, the situation has to be understood with due attention for the context, in order to be able to estimate effects of changes in the flow regime for the well-being of different groups of people. However, in a restoration case, where degradation has already taken place, the effects on peoples' well-being can be measured directly. An understanding of the context will remain useful in a restoration case to explain the different impacts on different groups of people.

A second difference between the two situations is that when ecosystem degradation has occurred, people may have found alternatives or migrated to other areas. Because of these adaptations, the flow regime, which resulted in the natural situation in certain well-being values related to the river ecosystem, may not necessarily have the same effect in a restored situation. People may not be willing to involve again in the same activities which they previously lost. Perhaps the infrastructure resulting in the degradation of the ecosystem has brought them welfare through irrigated agriculture and resulted in a higher standard of living. Therefore, restoration studies require additional questions on the preferences of the population.

This activity can be carried out in focus groups or as a questionnaire survey. Group discussions take less time, but require skilled facilitators. Questionnaire surveys require skills as well, but mostly in designing the questions, and ensuring that the interviewers understand these well. For a moderately skilled interviewer prepared interviews will be easier to conduct than group discussions. Questionnaire surveys have both advantages and disadvantages. Advantages are that they are relatively easy to teach others to do, although there is always the risk that the interviewers or the interviewees misunderstand the question. Such misunderstanding is hard to deduct from the filled-out answers, because in questionnaires mainly fixed answers (multiple choice) are used. Because of the fixed answers, analysis of the data is relatively easy and straightforward, and gives an impression of how many people in a certain group experience certain things. Questionnaire surveys should in fact only be used to quantify the relationships and confirm relationships which are already rather well understood. Such an understanding will follow from the conduction of group discussions in the first part of the fieldwork.

When the fieldwork is based on the hydrological situations identified by the population, all questions, for example on income from different activities, should preferably be answered for each of these situations. In conservation situations, people may not have been able to identify many different situations. In such a case, the focus is on the current situation and it will be useful to obtain information on how much of what goods and services are required to fulfil certain well-being needs. With an understanding of the context the impacts of deviations from the current situation should be estimated for the different stakeholder groups.

Data from statistics, such as fish yields or incidence of particular diseases will be of great help to quantify the link between ecosystem and well-being. Either the availability of such data may omit the need for collecting certain primary field data, or the data can be used to verify the collected field data.

6.3.5 Assessment of relationships
For all possible flow regimes and ecosystem conditions the related well-being value needs to be calculated. Depending on whether is dealt with a conservation or a restoration case, on the identified indicators and on the availability of data, it can be the most logical approach to derive knowledge rules or to directly base the result on measured well-being values.

Values based on measured values for various situations

In a restoration case, or a conservation case in which people have experienced various hydrological situations, it will be possible to assess well-being values for a range of different situations. This means that human well-being values can be coupled directly to these situations. It can be expected, however, that based on ecological analysis more situations will be identified than recognised by the local communities. This was the case in Sistan, where 18 different combinations of hydrological characteristics with distinct ecosystem conditions were identified by the ecologists, while the population identified only 3 situations. In this situation it is proposed to identify with which of the 18 situations the identified 3 situations correspond. For the remaining situations a well-being value can be derived based on the well-being values that are known. This approach is illustrated in Figure 6.1.

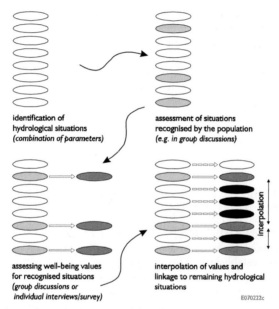

Figure 6.1 Steps to base values on values assessed for a limited number of situations

The advantage of this approach is that exactly how hydrology and ecology contribute to socio-economic values does not need to be understood. A disadvantage is that it is difficult to assess which of the possible hydrological situations identified by the ecologists are the ones understood by the population. Also, interpolation is unlikely to be linear and is therefore difficult without a good understanding of the system. Moreover, only values in between the recognised situations can be assessed. Preferably, absolute values for each of the identified situations are used in this approach, which is especially difficult for income. To deal with the situations for which the discussed approach is not possible, it is proposed to work with knowledge rules.

Knowledge rules

Knowledge rules can be defined for the calculation of human well-being values based on either a certain ecosystem condition or a flow regime characteristic. Especially in conservation cases only the value in the current situation can be assessed. Together with the understanding of the context a knowledge rule can be derived to estimate impacts of flow regime changes. A disadvantage is that an explicit understanding of the relationship is required which is difficult in complex situations and which requires a lot of information.

6.3.6 Description of possible second order impacts

Besides the quantified assessment of first order impacts, attention can and should be paid to second order impacts. This can be done in a qualitative way, but perhaps in the future this can be extended into a quantitative approach. The question is what will happen as a result of the first order impacts. Here again, it is important whether the changes have already taken place or are planned for the future. In case of future plans, people can be asked how they will respond to changes. In case of already experienced impacts, it would be interesting to discuss the various changes in the society, how people perceive these changes and how they explain the changes. Individual stories of impacts are useful to paint a picture, especially in combination with quantified impacts on people including their numbers. Once a good understanding is obtained about the different second order impacts, additional work can be done on the quantification. The second order impacts may feedback on the direct impacts of income and food, health and perception and experience, in either a positive or a negative way.

6.4 Step 3: assessing the relationship between the river ecosystem and the local flow regime

6.4.1 Description and envisaged results

In this step the relationship between the local flow regime and the ecosystem condition will be established. This is mainly a task for ecologists of various specialisations (e.g. in fish, birds, macrofauna, vegetation) together with hydrologists. However, close co-operation with social scientists is required to ensure that the step will yield results for the relevant goods and services in the desired format. Also, local knowledge on flow regime-ecosystem relationships and the occurrence of extreme or deviating situations can be a useful addition to the available data (Thompson & Polet, 2000). Similar to the previous step, the results of this step are either knowledge rules or tables or databases with ecosystem condition in terms of relevant goods and services for all possible hydrological situations.

6.4.2 Activities and considerations

The following activities are based on the DRIFT approach. Advantages of the DRIFT approach are that it can easily deal with uncertainty and that the method provides quantified results. The reader is referred to the DRIFT literature for further explanation (Brown & King, 2000; King et al., 2003). DRIFT is an approach which consists of various steps, some of which are overlapping with other steps in the application of the conceptual model. For this particular part of the approach developed in this thesis, the following steps of the DRIFT approach are the most relevant:

- Analysis of the natural flow regime, and identification of relevant hydrological parameters and thresholds.
- Estimation of ecosystem condition relative to the natural condition for each combination of parameters and thresholds (all possible hydrological situations). To include uncertainty, a range of change can be given, instead of a single value.

The context, such as over-exploitation or pollution, can be directly translated into relevant hydrological characteristics, or can be stated as a prerequisite for the established relationship to be valid at all.

6.5 Step 4: assessing the relationship between the local flow regime and the upstream flow regime

6.5.1 Description and envisaged results

In this step, a relationship needs to be established between an upstream flow regime and the local flow regime at the locations for which ecosystem condition and related well-being will be assessed. The tool used for this step will in most cases be a model, either a water balance model, which is the type of model most often used in comprehensive IWRM analyses, or a hydrological or hydrodynamic model. This step will generally be a combined task of various specialists: hydrologists provide inflow series, hydraulic engineers provide information on structures and reservoir operation, agricultural specialists determine crop water use, etcetera. Wetland hydrology and hydrodynamics should be part of this assessment. Co-operation with ecologists is required to understand the required output parameters, location and timing of output as well as the relevant time-step in the simulations. The context, local abstractions and management which affect the relationship between up- and downstream flow regime will generally be included in such a model. Operation of local infrastructure or adjustments made by the local population changes the relationship between external and internal flow regimes. When such situations are envisaged this should whenever possible be taken into account when considering the results. If permanent, the relationship should be adjusted, if temporary, different possible situation should be taken into consideration.

6.5.2 Activities and considerations

Various tools are available, of which the RIBASIM model, used in the Hamoun wetlands case study, is an example of a water balance model. Most water balance models which are currently available provide the same feature for including environmental flows: a minimum flow node. Such a node requires as input a pre-defined series of monthly minimum flows. A difficulty in linking this to environmental flows is the fact that environmental flows do not require an exact volume per time step, but rather need a certain flow characteristic to occur within a certain time period at the timescale of seasons or years. For example, a flood is required in a certain season, but it does not matter much during which month. Also droughts or large floods may be required on an interannual basis, which can not be captured in monthly series which are the same for each year. The best solution is perhaps to assess ecosystem impacts, and subsequently well-being impacts, through post-processing of the simulation results without defining fixed minimum flow requirements. The assessment was carried out in this way for the case of the Hamoun wetlands. For further refinement two additional analyses can be carried out using DRIFT solver in the DRIFT database. With this optimalisation module the best distribution over the year of a certain volume of water can be calcuated. This can help to further refine the flow requirements while assessing the impact on other sectors.

Hydrodynamic models generally do not include modules for water demand and management. Abstractions can be included as fixed time series of lateral outflow. Also for this type of models ecosystem and human well-being impacts need to be assessed through post-processing of the simulation results.

6.6 Step 5: estimating impacts of water resources management measures on the well-being of the identified stakeholder groups

6.6.1 Description and envisaged results

In this step, the actual assessment of the human well-being impacts for the different groups of stakeholders as a result of changed water resources management will take place, based on the established relationships in the previous steps. The result is typically a score card which contains the impacts in terms of the identified human well-being indicators for all selected strategies and scenarios. It is useful to include in such a scorecard also the resulting values for the hydrological parameters and for the ecosystem condition. The information resulting from this step is meant as input in the decision-making process.

6.6.2 Activities and considerations

Activities in this step consist of translating various water resources management strategies into ecosystem condition and human well-being impacts, based on the established relationships. Suggested is in the presentation of the results to neither aggregate the different well-being indicators into an overall well-being value, nor the impacts on the different stakeholder groups into one equity value. The information is not meant to provide an answer on what the single best strategy is, but serves as information to support negotiations and trade-offs. Not-aggregated data will serve this purpose better.

6.7 Suggested additional steps

If time and resources allow, two additional steps can be carried out. The first is a discussion of the results with the potentially affected communities. The second consists of monitoring livelihood impacts, as well as compliance with agreed flow regimes, after implementation.

6.7.1 Discussion with local communities

A discussion of the estimated results of potential strategies with the population would be highly useful, for the following reasons:

- Control of and feedback on the analysis, leading to a correction of established links and resulting impacts on human well-being.
- Insight in acceptability and preference of mitigation and compensation measures.

- Insight in second order effects if not yet assessed: if the estimated effects will indeed occur in a certain way, for example with reduced income, harsh living conditions, or alternate good and bad years, how will the population respond? This can be added as indicator 'second order effects' on the scorecard.

This discussion is also useful in the light of the general IWRM process: people will better understand their situation, what is possible and what not. Such an understanding is useful if the population (or their representatives) are to be involved in participatory decision-making or negotiation. A prerequisite for such a discussion with the population is that the authorities are indeed willing to take measures and to listen to the opinion of the population. Otherwise, unnecessary expectations may be raised, which should be avoided.

6.7.2 Monitoring and adaptive management

Despite a thorough assessment of human well-being impacts, actual responses of people to changes in their environment are difficult to predict. It is important that during the planning stage concrete agreements will be made on the desired level of well-being and ecosystem condition. After implementation of the water resources management strategy the changes in well-being and ecosystem should be monitored. When effects are different than agreed upon, the management strategies should be adjusted. This requires agreements before project implementation on how monitoring results will be used in adaptive management. Another reason for not arriving at the agreed well-being and ecosystem condition could be non-compliance with the agreed management strategies, especially where it concerns operation of reservoirs. Therefore, not only well-being and ecosystem condition should be monitored, but also the operation of the water resources system (Brown & Watson, 2006). This information can be used to check the accurateness of the established links between hydrology, ecosystem condition and human well-being. For the monitoring of well-being, the same activities as described in step 2 can be repeated on a regular basis. Another solution would be to select key-informants and to contact them frequently.

6.8 Conclusions

This chapter discussed the differences between the two case studies for each of the steps taken to apply the conceptual model and the lessons learned from the application of the conceptual model. Comparison of the two cases revealed for steps 1 and 2 differences in data collection and analysis methods. One of these differences was the climatological situation. In arid areas (Hamoun wetlands), the boundaries of the river ecosystem can be more clearly defined than in wet areas (Surma-Kushiyara rivers). In arid areas, it will therefore in general be easier to identify the people who depend on the river ecosystem. A second difference is the difference between conservation (before intervention) and restoration (after intervention) cases. When interventions have already taken place, information on impacts can be collected more directly than in situation where changes have not yet taken place.

Analysis of steps 3 and 4 revealed that linking human well-being to the results of environmental flow assessment methods and water allocation methods can indeed be done. However, in both cases additional post-processing of the results obtained with the environmental flow assessment and water allocation methods proved necessary.

The comparison of the case studies does not reveal a need for adjustments of the conceptual model. The conceptual model and stepwise approach are general and have proven to provide a structured framework for the analysis of the importance of the river flow regime for the well-being of various stakeholder groups. The lessons learned through the case studies can serve as guidelines to support future application of the conceptual model for the assessment of human well-being related to environmental flows.

In many situations there may not be sufficient time or resources to conduct a full assessment as discussed in this chapter, and in a technical assessment there may not always be a willingness to spend time and efforts on social studies. However, it is strongly suggested to make at least an inventory of the potentially affected people, with their link to the ecosystem, the components of their well-being most likely affected and the extent to which these components will be affected. With additional time, group discussions can be conducted to elaborate this information. With these steps the assessment is mainly qualitative. To carry out a quantitative assessment, the relationships between human well-being indicators and water resources management need to be quantified which involves all remaining activities of steps 2 through 5. A distinction can be made between focusing on the first order impacts, or to include also secondary impacts in either a qualitative or also a quantitative way. Depending on the scope of the project, additional time and resources can be used to discuss the results with the stakeholders and to monitor the impacts after implementation of the selected strategy.

The list of steps may unrightfully suggest that social scientists start and ecologists and hydrologists only come in later. It is of utmost importance that all specialists are involved from the beginning and exchange the relevant information. Social scientists need information from hydrologists and ecologists to identify the stakeholders, to make meaningful questionnaires and to interpret the data collected during group discussions or interviews. Hydrologists and ecologists at the other hand need the information from the social scientists to ensure that their results will match and can benefit from the local knowledge and experiences.

7 Discussion and Conclusions

7.1 Introduction

This chapter discusses the final achievements of the presented research in terms of the objective and research questions as well as the contribution to science and society. The objective for this thesis was formulated as:

> *To develop an approach for the assessment of human well-being values and social equity related to environmental flows for application in Integrated Water Resources Management.*

Three research questions were formulated:

1. What are the approaches currently used in environmental flow assessments and Integrated Water Resources Management, and how is or can human well-being be linked to these approaches?
2. What is the conceptual relationship between river flows, river ecosystems and human well-being and social equity, and how does this conceptual relationship fit in the approaches in environmental flow assessments and Integrated Water Resources Management?
3. How can the conceptual relationships be quantified in practical applications?

Through an assessment of the human well-being value of environmental flows this thesis hopes to contribute to informed and equitable water resources management, and in this way to contribute to the protection of both the river ecosystem and the well-being of the people depending on this ecosystem.

The following sections first answers the research questions, and subsequently discusses the contribution to science and society. The chapter concludes with an overall conclusion for this thesis.

7.2 Research question 1: current approaches

Chapter two discussed literature on both environmental flows and Integrated Water Resources Management, to obtain insight in how approaches used in the two fields can be connected, and how human well-being is or can be dealt with. The following conclusions can be drawn:

- To include human well-being related to environmental flows in IWRM the type of environmental flow assessment methods referred to as 'holistic' and 'scenario-based' are the most appropriate. Holistic methods claim to include all parts of the natural ecosystem related to river flows. Because people use various goods and services of the river ecosystem, it is required to use such a holistic approach to assess the relationship between river ecosystems and human well-being. Scenario-based approaches estimate impacts of changed flow regimes on various components of the ecosystem, opposed to the objective-based approaches which determine a desired ecosystem condition upfront and subsequently derive the flow regime required to maintain this specified condition. A problem with the objective-based approaches is that it is difficult to identify what the desired ecosystem condition is. Moreover, the cause-effect approach of the scenario-based methods is similar to the analysis of impacts of water resources management strategies in the systems analysis approach of IWRM. It can be concluded that the holistic scenario-based approaches are the most suitable to include human well-being related to environmental flows in IWRM analysis.

- Human well-being related to environmental flows should be assessed as one of the criteria to evaluate the functioning of the water resources system under various strategies and scenarios. Currently, IWRM refers to three criteria: economic efficiency, social equity and environmental sustainability. However, IWRM does not specify explicit criteria for human well-being or social equity linked to natural ecosystems. In water allocation literature, the sectors of public water supply and irrigated agriculture are recognized as important to sustain human well-being, whereas allocation of water to nature is mainly appreciated to protect the resource base for future generations. Informed water resources management decision-making requires quantification of the links between all water use sectors and all water resources management criteria. It can be concluded that there is a need to broaden the social equity criterion to include the link with the ecosystem and to conceptualise this link.

- Current holistic scenario-based approaches do not provide quantifiable transparent relationships to assess the importance of river ecosystems for human well-being. Such quantified relationships are required to ensure proper consideration of the importance of river ecosystems for human well-being and social equity in IWRM decision-making.

7.3 Research question 2: conceptual relationship and link with environmental flow assessments and integrated water resources management

A conceptual model of the relationship between river flows, river ecosystems and human well-being has been constructed based on literature, as well as through incorporation of the experiences of two case studies. The conceptual model has been constructed in such a way that the main gaps in current IWRM and EF approaches are reduced: the identification of quantifiable indicators for human well-being and the enabling of the assessment of the different impacts for different groups of people. To apply the conceptual model five steps are identified, which are linked to the steps of the systems analysis approach of IWRM:

1. identifying stakeholder groups;
2. assessing the relationship between human well-being and the river ecosystem;
3. assessing the relationship between the river ecosystem and the local flow regime;

4. assessing the relationship between the local flow regime and the upstream flow regime; and
5. estimating impacts of water resources management measures on the well-being of the identified stakeholder groups.

The following main conclusions can be drawn:

- To measure changes in human well-being, changes in human well-being values need to be assessed for three well-being components: income & food, health and perception & experience. Operational indicators for each of these main components of human well-being need to be defined according to the particular situation. Because the socio-economic system is complex and dynamic, changes in these three components of well-being are likely to impact other components of well-being such as independence, social structure and other psycho-social factors. These components of well-being are in this thesis referred to as second order, because they are indirectly affected by the changes in the river ecosystem. As a result of second order impacts, the first order components can change further, in a positive or negative direction.
- To understand not only the link with, but also the importance of, a change in the ecosystem for human well-being, it is necessary to explicitly consider the context, as proposed in the conceptual model. Through consideration of the context, it is possible to assess for example whether the identified groups of people have indeed access to the goods and services provided by the river ecosystem, and if so, whether they have alternative resources at their disposal.
- To understand whether water resources management strategies lead to social equity, it is important to identify distinct groups of people who are likely to experience different impacts. Through a quantification of the importance of changes in the river ecosystem for various groups of people, the contribution to social equity of alternative water resources management strategies can be assessed.
- This thesis aimed at extending the existing environmental flow and IWRM approaches with a link to human well-being and social equity. To assess the human well-being impacts of changed flow regimes, not only the ecosystem – well-being links, but also the links between upstream and local flow regimes, ecosystem condition and goods and services needed to be included in the conceptual model. Of these, the links between local flow regimes and ecosystem condition are the topic of most of the available environmental flow methods, while the link between upstream and local flow regime is the covered in the field of hydraulics and hydrologic routing. Where possible key features of existing approaches were included in the model.

7.4 Research question 3: quantification in practical applications

The approach was applied in, and partly developed based on, two case studies. The two cases were very different in many ways, which has shown that the approach is generally applicable, but that different situations may lead to different well-being criteria, different data collection methods, different tools to be used, and different results. In Chapter 6, lessons learned in the case studies for carrying out the five steps identified in Chapter 3 were discussed. The major conclusions for a good practical assessment can be summarised as follows:

- It is important to define with the stakeholders what is meant by flood or drought, and what different situations can be distinguished. Only then further group discussions or questions can be meaningful. This also implies that a questionnaire alone is not likely to yield satisfying results, because questions may be easily interpreted in a different way.
- It is useful to distinguish situations where the impacts of future changes need to be assessed (referred to as conservation situation), and where changes have taken place already (referred to as restoration situation). In the restoration situation people have experience with different conditions of the ecosystems, which enables a direct assessment of well-being values in each of the situations. In this case it is also possible that people have found alternative ways of living or have migrated, which means that restoring the ecosystem to a previous condition does not have to imply a similar use and well-being value of the ecosystem.
- Information on the context needs to be collected together with information on each of the links. The information on the context will either become part of the actual quantified relationship, or will be a prerequisite for certain relationships to be valid. In restoration situations where human well-being impacts of ecosystem changes can be measured directly, the context can be used to explain why different groups of people experience different changes in their well-being.
- Assessing impacts of changed flow regimes on ecosystems focuses on the average condition of an ecosystem over a long period of time. Floods and droughts may be required to maintain this condition. For people, however, such extremes should generally be avoided. To assess impacts for people the focus needs to be on the availability of goods and services per year, or similarly on the frequency of years with shortages.
- Practical assessment requires various specialists to work together. Information exchange between social scientists, ecologists and hydrologist is required in all steps of the conceptual model and will be mutually beneficial.
- Because of the complexity and inherent uncertainty in predicting social impacts, discussing the outcomes of the analysis with the local population and monitoring the actual impact after implementation of new strategies is highly recommended.

7.5 Contribution to science and society

The three research questions have been discussed. But what has this thesis contributed to science and society? From the discussion in Chapter 1 it has become clear that with regard to science, the study should contribute to both the fields of environmental flows and of Integrated Water Resources Management. With regard to society it is the contribution to the protection of nature and of the poor and ignored communities. The contribution of this thesis to these four fields and issues are discussed in this section.

7.5.1 Contribution to Integrated Water Resources Management

In IWRM literature the allocation of water to various water demand sectors is linked to decision-making criteria. This thesis has provided more insight in one of the links which has not received much attention before: the link between nature as water demand sector and the criterion of social equity. For this purpose social equity is divided into a number of well-being components: income & food, health, perception & experience and second order components. For each of these components indicators can be identified, which can be linked to the decision-making criteria.

Assessment of these indicators and links with ecosystems and water allocation in the case study on the Hamoun wetlands (Chapter 5) lead to additional difference between alternative water resources management strategies. This means that other decisions could be taken when human well-being is explicitly taken into consideration. This shows that the generated information is not redundant but provides useful additional information, and confirms the need for the research discussed in this thesis.

The method developed in this thesis provides a structured approach to carry out a quantified assessment of human well-being related to river ecosystems and river flows, and this way has extended the systems analysis approach often applied in IWRM. Quantification is important to give ecosystems and related human well-being due consideration, similar to the quantified benefits from allocating water to agriculture or other uses. This approach can support researchers and consultants in the field of IWRM. And subsequently, this may enhance informed decision-making with regard to water resources management.

Participatory decision-making is nowadays becoming more and more common in water resources management. Meaningful participation of the public requires information about the likeliness of relevant impacts on local communities. Through the information generated, the approach developed in this thesis can support public participation.

7.5.2 Contribution to environmental flows

A structured approach to assess the importance of flows for people as part of environmental flow assessments was identified by various authors as a missing part in current environmental flow assessment methods (Chapter 1). This thesis presents an approach to conduct such an assessment in a quantified way, and through this contributes to the field of environmental flows science. For applying this approach it is of utmost importance that there is a continuous co-operation between hydrologists, ecologists and social scientists during the course of the assessment.

Furthermore, the thesis has further clarified the links between current approaches in both the fields of environmental flows and IWRM, and through this contributed to both fields.

7.5.3 Contribution to the protection of nature

The approach developed focuses on quantification, since this is the approach used in systems analysis in IWRM. This quantification is believed to contribute to the protection of nature, because when decision-makers base their decision on quantified scores in a score-card, not being able to quantify the importance of nature on human well-being implies the risk that these values are ignored. Of course, next to the assessment of the importance of ecosystems for human well-being, the economic value of ecosystems as well as the value for biodiversity conservation need to be assessed.

Quantification also has its pitfalls. In the presented approach the most direct and more tangible impacts are the first to be quantified. Second order impacts are not (yet) quantified. Also, as Einstein once said: "not everything that can be counted counts and not everything that counts can be counted". A qualitative assessment may be more suitable to capture the less tangible as well as the second order impacts. Yet, even though quantified values may not represent the full value for the local communities, they are useful to compare impacts of various strategies.

Another pitfall is that the focus of decision-making authorities is naturally on the communities under their responsibility. This may imply that the importance of the ecosystem for other people is not assessed. Not only does this mean that not the full value of the ecosystem is accounted for, but also that opportunities for wetland management may be missed. When people who live at (large) distances from a particular ecosystem, value this ecosystem for its intrinsic or recreational value, or for its contribution to their climate and living conditions, they may be willing to pay for the conservation of this area. It would be unfair that poor local communities are denied the opportunities of irrigated agriculture in order to conserve nature which benefits other people.

Decision-makers are often triggered by two things: money and people to vote for them. Therefore, despite the pitfalls mentioned in the above, quantifying the importance of ecosystems for the well-being of people may yield better results for the protection of nature than describing the effect on nature itself.

7.5.4 Contribution to the protection of local communities

The socio-economic system is highly complex. Poverty has many aspects, and water can influence a number of these aspects to a certain extent. This thesis has tried to provide insight in the contribution of flow-dependent ecosystems to human well-being, and in this way hopes to contribute to the protection of this part of human well-being of the local communities. Yet, due to the complexity and dynamic character of the socio-economic system, there is a risk that not all impacts are considered or not quantified to the correct extent. Careful monitoring in combination with adaptive management plans is therefore pertinent.

This study focuses on the importance of river ecosystems for local communities, in order to protect the interests of these communities. This should not be mistaken for advocating against development. Literature has shown that economic development in general has improved the living standards of the population (Barker *et al.*, 2000; Rijsberman, 2003). According to various authors, developments are necessary in order to feed and increase the well-being of the growing world population (Chambers, 1983 p 175; Scudder, 2005). The well-being of the people depending on the river ecosystem can protected in two ways: 1) through protecting the ecosystem goods and services that are important for the well-being of the local communities, and 2) through providing alternatives.

Both solutions, protecting the ecosystem or providing alternatives, require an assessment of how development projects will impact the ecosystem and the associated well-being of the communities. This thesis does not advocate a certain answer or claim that a certain amount of water should be allocated to the poor. What is important is that all impacts and the sharing of these benefits and losses are given due consideration in the decision-making process. It is to the quantification of the sharing of benefits and losses that this thesis aims to contribute. Since both solutions may be acceptable from the point of view of local communities, this implies that protecting the communities who depend on the natural ecosystem, does not necessarily lead to a protection of this ecosystem.

Whether an assessment of human well-being related to river ecosystems and water resources management will be conducted, largely depends on whether the decision-makers, government authorities or donor organisations ask for it. Consultants will normally not consider what is not in their Terms of Reference. However, in many situations project objectives could be formulated in broad terms, and leave room for consultants to decide what should be part of such an assessment. In such a situation it is important that consultants are aware of the importance of river flows for ecosystems and human well-being and of the methods to assess environmental flows and well-being impacts. This thesis is meant to provide a structured approach to be used in such an assessment, and this way hopes to facilitate the consideration of well-being impacts in water resources decision-making.

7.6 Final conclusion

The thesis provides an approach for assessing human well-being values of environmental flows as part of IWRM studies, consisting of a conceptual model together with a stepwise approach for the actual assessment. With this approach the thesis makes a contribution to the further operationalisation of the concepts of IWRM and environmental flows. The approach presented in this thesis generates more comprehensive and more socially-relevant information to decision-makers, which is essential to enhance social equity in IWRM.

7.7 Future research

This thesis describes one of the first approaches to include environmental flows and human well-being in an IWRM analysis. The research for this thesis has been done from a water resources management point of view. Now that links between human well-being and IWRM are better understood, it will be interesting to investigate to what extent methods used in various fields of social science can be used to improve and elaborate the identification of stakeholder groups and the assessment of the relationships between the well-being of the stakeholders and the river ecosystem and river flow regime. This requires further elaboration of the second order impacts as well as monitoring of actual well-being impacts after interventions in upstream river stretches have taken place.

The approach described in this thesis has been applied in two case studies. However, the results of these two case studies have not been used in actual decision-making processes yet. It will be useful to learn to what extent the additional generated information will indeed lead to better-informed decision-making and whether different decisions are or would be taken with information on human well-being values and impacts. Analysing the decision-making processes will provide insight in the required scope and detail of information on human well-being impacts of flow regime and ecosystem changes.

Related to the decision-making process is the general institutional setting. In this thesis the focus has been on the relationship between the Natural Resources System (river flow regime and river ecosystem) and the Socio-Economic System (the stakeholders and their well-being). However, to improve the understanding of how information on this relationship can influence water resources management decisions, the links with the Administrative and Institutional System, should be further considered.

References

References

Acreman, M. C., F. A. K. Farquharson, M. P. McCartney, C. Sullivan, K. Campbell, N. Hodgson, J. Morton, D. Smith, M. Birley, D. Knott, J. Lazenby, R. Wingfield & E. B. Barbier, 2000. Managed flood releases from reservoirs: issues and guidance. Centre for Ecology and Hydrology, Wallingford, UK.

Adams, A., 1999. Social Impacts of an African Dam: Equity and Distributional Issues in the Senegal River Valley. World Commission on Dams, Cape Town.

Adams, W., 2000. Downstream Impacts of Dams. World Commission on Dams, Cape Town.

Arthington, A. H., J. L. Rall, M. J. Kennard & B. J. Pusey, 2003. Environmental Flow Requirements of Fish in Lesotho Rivers using the DRIFT methodology, River Research and Applications 19: 641-666.

Ashley, C.& D. Carney, 1999. Sustainable livelihoods: Lessons from early experience. DFID, London.

Barbier, E. B., M. C. Acreman & D. Knowler, 1997. Economic valuation of wetlands; a guide for policy makers and planners. Ramsar Convention Bureau; Department of Environmental Economics and Environmental Management, University of York; Institute of Hydrology; IUCN-The World Conservation Union.

Barbier, E. B.& J. R. Thompson, 1998. The Value of Water: Floodplain versus Large-scale Irrigation Benefits in Northern Nigeria, Ambio 27: 434-440.

Bari, M. F.& M. Marchand, 2006. Introducing Environmental Flow Assessment in Bangladesh: Multidisciplinary Collaborative Research. Final Technical Report, BUET-DUT Linkage Project phase III. Capacity Building in the Field of Water Resources Engineering and Management in Bangladesh. BUET-DUT Linkage Project.

Barker, R., B. Van Koppen & T. Shah, 2000. A Global Perspective on Water Scarcity/Poverty; Achievements and Challenges for Water Resource Management. International Water Management Institute, Colombo, Sri Lanka.

BBS, 1996. Bangladesh Population Census 1991. Zila: Sylhet. Bangladesh Bureau of Statistics.

BBS, 2001. Census of Agriculture-1996. Zila: Sylhet. Bangladesh Bureau of Statistics.

References

Brown, C. & J. M. King, 2000. Environmental Flow Assessments for Rivers: A Summary of the DRIFT Process. Southern Waters, Mowbray, South Africa: 1-27.

Brown, C. & J. M. King, 2003. Environmental Flows: Concepts and Methods. In Davis, R. & R. Hirji (eds), World Bank, Washington: 1-30.

Brown, C. A. & P. L. Watson, 2006. Decision Support Systems for Environmental Flows: Lessons from Southern Africa.Enhancing equitable livelihood benefits of dams using decision support systems. Adama/Nazareth, Ethiopia.

Bunn, S. E. & A. H. Arthington, 2002. Basic Principles and Ecological Consequences of Altered Flow Regimes for Aquatic Biodiversity, Environmental Management 30: 492-507.

Cernea, M. M., 2000. Risks, safeguards, and reconstruction: a model for population displacement and resettlement. In Cernea, M. M. & C. McDowell (eds), Risks and Reconstruction: Experiences of Resettlers and Refugees. The World Bank, Washington D.C.: 11-55.

Chambers, R., 1983. Rural Development - Putting the Last First. Pearson, Essex.

Chambers, R., 1995. Poverty and livelihoods: whose reality counts?, Environment and Urbanization 7: 173-204.

Chambers, R. & R. G. Conway, 1991. Sustainable rural livelihoods: practical concepts for the 21st century. Institute of Development Studies, University of Sussex, Sussex.

Cosgrove, W. & F. R. Rijsberman, 2000. World Water Vision; Making Water Everybody's Business. Earthscan, London.

Costanza, R., R. d'Arge, R. De Groot, S. Farber, M. Grasso, B. Hannon, K. Limburg, S. Naeem, R. V. O'Neill, J. Paruelo, R. G. Raskin, P. Sutton & M. Van den Belt, 1997. The value of the world's ecosystem services and natural capital, Nature 387: 253-260.

De Graaf, G., B. Born, A. M. Kamal Uddin & F. Marttin, 2001. Floods, Fish and Fishermen. Eight Years Experiences with Flood Plain Fisheries, Fish Migration, Fisheries Modelling and Fish Bio Diversity in the Compartmentalization Pilot Project, Bangladesh. The University Press Limited, Dhaka, Bangladesh.

De Groot, R. S., 1992a. Functions of Nature; Evaluation of nature in environmental planning, management and decision making. Wolters-Noordhoff.

De Groot, W. T., 1992b. Environmental Science Theory; concepts and method in a one-world, problem-oriented paradigm. Elsevier, Amsterdam.

DFID, 2001. Biodiversity - a crucial issue for the world's poorest.

DHI, 2003. MIKE BASIN: a tool for river basin planning and management. User Manual for MIKE BASIN 2003. Danish Hydraulic Institute, Horsholm, Denmark.

Douglas, A. J.& R. L. Johnson, 1991. Aquatic Habitat Measurement and Valuation: Imputing Social Benefits to Instream Flow Levels, Journal of Environmental Management 267-280.

Drijver, C. A.& M. Marchand, 1986. Taming the floods: environmental aspects of floodplain development in Africa. UNESCO, 13-22.

Dugan, P., 1990. Wetland Conservation. A Review of Current Issues and Required Action. IUCN, Gland, Switzerland: 1-96.

Dunbar, M. J., M. C. Acreman & S. Kirk, 2004. Environmental Flow Setting in England and Wales: Strategies for Managing Abstraction in Catchments, Water and Environment Journal 18: 5-10.

Dunbar, M. J., A. Gustard, M. C. Acreman & C. R. N. Elliot, 1998. Overseas appraoches to setting River Flow Objectives. Environment Agency, Bristol: 1-83.

Emerton, L.& E. Bos, 2004. Value: counting ecosystems as an economic part of water infrastructure. IUCN, Gland.

ESCAP, 1992. Towards and environmentally sound and sustainable development of water resources in Asia and the Pacific. United Nations, New York.

ESCAP, 2000. Principles and practices of water allocation among water-use sectors. United Nations, New York.

Fiselier, J. L., 1990. Living off the Floods: Strategies for the integration of conservation and sustainable resource utilization in floodplains. EDWIN (Environmental Database on Wetland Interventions, Leiden.

Glantz, M. H.& I. S. Zonn, 2005. The Aral Sea: Water, climate, and environmental change in Central Asia. World Meteorological Organization.

Gram, S., L. P. Kvist & A. Caseres, 2001. The Economic Importance of Products Extracted from Amazonian Flood Plain Forests, Ambio 30: 365-368.

GWP, 2000. Integrated Water Resources Management. Global Water Partnership, Denmark.

GWP, 2003. Poverty Reduction and IWRM. Global Water Partnership Technical Committe.

Hermans, L., 2005. Actor analysis for water resources management. Putting the promise into practice.

Hickey, J. T.& G. E. Diaz, 1999. From flow to fish to dollars: An integrated approach to water allocation, Journal of the American Water Resources Association 35: 1053-1067.

Horowitz, M. M., 1991. Victims Upstream and Down, Journal of Refugee Studies 4: 164-181.

Hussain, I.& M. Giordano, 2004. Water and Poverty Linkages; Case Studies from Nepal, Pakistan and Sri Lanka. IWMI, Colombo.

Iran Statistics Institute, 2002. Unknown.

Iran Statistics Institute, Z. D., 1966. Statistics on Population and Residence 1966, Zabol. Iran Statistics Institute, Zabol, Iran.

Iran Statistics Institute, Z. D., 1976. Statistics on Population and Residence 1976, Zabol. Iran Statistics Institute, Zabol, Iran.

Iran Statistics Institute, Z. D., 1986. Statistics on Population and Residence 1986, Zabol. Iran Statistics Institute, Zabol, Iran.

Iran Statistics Institute, Z. D., 1996. Statistics on Population and Residence 1996, Zabol. Iran Statistics Institute, Zabol, Iran.

IUCN, 2000. Vision for Water and Nature. A World Strategy for conservation and Sustainble Management of Water Resources in the 21st Century. Switserland, Gland.

Jobin, W. R., 1999. Dams and Disease: Ecological design and health impacts of large dams, canals and irrigation systems. Routledge, London and New York.

Jowett, I. G., 1997. Instream Flow Methods: A comparison of approaches, Regulated Rivers: Research & Management 13: 115-127.

Junk, W. J., P. B. Bayley & R. E. Sparks, 1989. The Flood Pulse Concept in River-Floodplain Systems, Can. Spec. Publ. Fish. Aquat. Sci. 106: 110-127.

Karim, K., M. E. Gubbels & I. C. Goulter, 1995. Review of determination of instream flow requirements with special application to Australia, Water Resources Bulletin 31: 1063-1077.

King, J. M., C. Brown & H. Sabet, 2003. A scenario-based holistic approach to envrionmental flow assessments for rivers, River Research and Applications 619-639.

King, J. M., R. E. Tharme & C. Brown, 1999. Definition and Implementation of Instream Flows. World Commision on Dams, Cape Town: 1-87.

King, J. M., R. E. Tharme & M. S. De Villiers, 2000. Environmental Flow Assessments for Rivers: Manual for the Building Block Methodology. Freshwater Research Unit, University of Cape Town, Cape Town, South Africa.

Klasen, S., 2002. The Costs and Benefits of Changing In-stream Flow Requirements (IFR) below the Phase 1 Structures of the Lesotho Highlands Water Project (LHWP). 1-33.

Konradsen, F., W. Van der Hoek, C. Perry & D. Renault, 1997. Water: where from, and for whom?, World Health Forum 18: 41-43.

Kwadijk, J. C. J.& F. L. M. Diermanse, 2006. Integrated Water Resources Management for the Sistan Closed Inland Delta, Iran. Annex B: Forecasting the Flow from Afghanistan. WL | Delft Hydraulics, Delft, The Netherlands.

Lok-Dessalien, R., 1999. Review of Poverty Concepts and Indicators. UNDP.

Loomis, J. B., 1998. Estimating the public's values for instream flow: Economic techniques and dollar values, Journal of the American Water Resources Association 34: 1007-1014.

Loth, P., 2004. The return of the water: Restoring the Waza Logone floodplain in Cameroon. In Loth, P. (ed), IUCN, Gland.

Loucks, D. P.& E. Van Beek, 2006. Water Resources Systems Planning and Management; An Introduction to Methods, Models and Applications. UNESCO.

Mainka, S., J. McNeely & B. Jackson, 2005. Depend on Nature; Ecosystem Services supporting Human Livelihoods. IUCN.

Mansoori, J., 1994. The Hamoun Wildlife Refuge. Max Kasparek, Heidelberg: 1-57.

Marchand, M., 1987. The productivity of African floodplains, Intern. J. Environmental Studies 29: 201-211.

Marchand, M.& F. H. Toornstra, 1986. Ecologische richtlijnen voor de ontwikkeling van stroomgebieden (Ecological guidelines for the development of river basins).Documentatie bij het advies Milieu en ontwikkelingssamenwerking (Documentation by the advice Environment and development cooperation). Commissie Ecologie en Ontwikkelingssamenwerking.

Maslow, A., 1954. Motivation and Personality. Harper & Row, New York.

McCartney, M. P.& M. C. Acreman, 2001. Managed flood releases as an environmental mitigation option, Hydropower & Dams 74-80.

McCully, P., 2001. Silenced Rivers: The Ecology and Politics of Large Dams: Enlarged and Updated Edition. Zed Books; London.

McNeely, J., 1988. Economics and Biological Diversity: Developing and Using Economic Incentives to Conserve Biological Resources. International Union of the Conservation of Nature (IUCN), Gland.

Meijer, K. S., 2006. Integrated Water Resources Management for the Sistan Closed Inland Delta, Iran. Annex E: Socio-economic Valuation of Allocating Water to the Hamouns and to Agriculture. WL | Delft Hydraulics, Delft, The Netherlands.

Meijer, K. S.& E. Van Beek, 2006. Integrated Water Resources Management for the Sistan Closed Inland Delta. Annex C: Sistan Water Resources System - supply and demand. WL | Delft Hydraulics, Delft, The Netherlands.

Meinzen-Dick, R.& M. Bakker, 1999. Irrigation systems as multiple-use commons: Water use in Kirindi Oya, Sri Lanka, Agriculture and Human Values 16: 281-293.

References

Merrey, D. J., P. Drechsel, F. W. T. Penning de Vries & H. Sally, 2005. Integrating "livelihoods" into integrated water resources management: taking the integration paradigm to its logical next step for developing countries, Reg Environ Change 5: 197-204.

Millennium Ecosystem Assessment, 2005. Ecosystems and Human Well-being: Synthesis. Island Press, Washington, D.C..

Miser, H. J.& E. S. Quade, 1985. Handbook of Systems Analysis; overview of uses, procedures, applications, and practice. John Wiley & Sons.

Molden, D.& C. De Fraiture, 2004. Investing in Water for Food, Ecosystems and Livelihoods.

Molle, F.& P. Mollinga, 2003. Water poverty indicators: conceptual problems and policy issues, Water Policy 5: 529-544.

Moore, M., 2004. Perceptions and interpretations of Environmental Flows and implications for future water resource management; A Survey Study. Department of Water and Environmental Studies, Linkoping University, Sweden, 1-67.

Mostert, E., E. Van Beek, N. W. M. Bouman, E. Hey, H. H. G. Savenije & W. A. H. Thissen, 2000. River Basin Management and Planning. In Mostert, E. (ed), Riber Basin management; Proceedings of the International Workshop (The Hage, 27-29 October 1999). UNESCO, 24-55.

Mouafo, D., E. Fotsing, D. Sighomnou & L. Sigha, 2002. Dam, Environment and Regional Development: Case Study of the Logone Floodplain in Northern Cameroon, Water Resources Development 18: 209-219.

MRC, 2003. State of the Basin Report. Mekong River Commission, Phnom Penh, Cambodia.

Narayan, D., 1999. Can Anyone Hear Us? Voices From 47 Countries. World Bank.

Nyambe, N.& C. Breen, 2002. Environmental Flows, Power relations and the use of river system resources. Cape Town: 1-13.

O'Keeffe, J., 1999. The South African experience in assessing and managing environmental flow requirements. Harare, Zimbabwe.

O'Keeffe, J.& D. Louw, 2000. Ecological Management Classes. In King, J. M., J. E. Tharme & M. S. De Villiers (eds), Water Research Commission, Pretoria: 125-132.

Olden, J. D.& N. L. Poff, 2003. Redundancy and the choice of hydrologic indices for characterizing streamflow regimes, River Research and Applications 19: 101-121.

Pearce, F., 2001. Bangladesh's arsenic poisoning: who is to blame?

Penning, W. E.& A. J. Beintema, 2006. Integrated Water Resources Management for the Sistan Closed Inland Delta, Iran. Annex D. Sistan Wetland Ecosystem - Functioning and Responses. WL | Delft Hydraulics, Delft, The Netherlands.

Pirot, J.-Y., P. J. Meynell & D. Elder, 2000. Ecosystem Management: Lessons from around the World. A Guide for Development and Conservation Practitioners. IUCN, Gland, Switzerland, and Cambridge, UK.

Poff, N. L., J. D. Allan, M. D. Bain, J. R. Karr, K. L. Prestegaard, B. D. Richter, R. E. Sparks & J. C. Stromberg, 1997. The Natural Flow Regime: A paradigm for river conservaton and restoration, BioScience 47: 769-784.

Pollard, S. R., 2000. Social use of riverine resources. In King, J. M., J. E. Tharme & M. S. De Villiers (eds), Water Research Commission, Pretoria: 95-115.

Pollard, S. R., 2002. Giving people a voice: Providing an environmental framework for the social assessment of riverine resource use in the Sabie River, South Africa. Cape Town: 1-13.

Pollard, S. R.& A. Simanowitz, 1997. Environmental flow requirements: A social dimension. In Pickford, J. (ed), WEDC, Lougborough: 397-400.

Postel, S.& B. D. Richter, 2003. Rivers for Life.

Richter, B. D., J. V. Baumgartner, R. Wigington & D. P. Braun, 1997. How much water does a river need?, Freshwater Biology 37: 231-249.

Rijsberman, F. R., 2003. Can development of water resources reduce poverty?, Water Policy 5: 399-412.

Rijsberman, F. R.& A. Mohammed, 2003. Water, food and environment: conflict or dialogue, Water Science and Technology 47: 53-62.

Rijsberman, F. R.& D. Molden, 2001. Balancing Water Uses: Water for Food and Water for Nature. In Secretariat of the International Conference on Freshwater - Bonn 2001 (ed).

Schuyt, K. D., 2005. Economic consequences of wetland degradation for local populations in Africa, Ecological Economics 53: 177-190.

Schuyt, K. D.& L. Brander, 2004. The Economic Values of the World's Wetlands. WWF; Gland/Amsterdam.

Scoones, I., 1998. Sustainable Rural Livelihoods: a Framework for Analysis. Institute of Development Studies, University of Sussex, Brighton.

Scudder, T., 2002. People dependence on environmental flows. Cape Town: 1-19.

Scudder, T., 2005. The Future of Large Dams; Dealing with Social, Environmental, Institutional and Political Costs. Earthscan, London; Sterling, VA.

SEI, 2001. User guide for WEAP-21. Stockholm Environment Institute, Tellus Institute, Boston, Massachusetts.

Sen, A., 1993. Capability and Well-Being. In Nussbaum, M. & A. Sen (eds), The Quality of Life. Clarendon Press, Oxford.

Silvius, M. J., M. Oneka & A. Verhagen, 2000. Wetlands: Lifeline for People at the Edge, Physics and Chemistry of the Earth 25: 645-652.

SLI& NHC, 1993a. Northeast Regional Water Management Plan. Appendix: Initial Environmental Evaluation. In SNC-LAVALIN International & Northwest Hydraulic Consultants Ltd. (eds), Bangladesh Water Development Board.

SLI& NHC, 1993b. Northeast Regional Water Management Plan. Main Report. In SNC-LAVALIN International & Northwest Hydraulic Consultants Ltd. (eds), Bangladesh Water Development Board, Dhaka, Bangladesh.

SLI& NHC, 1993c. Northeast Regional Water Management Plan. Surface Water Resources of the Northeast region. Preliminary Draft for discussion only. In SNC-LAVALIN International & Northwest Hydraulic Consultants Ltd. (eds), Bangladesh Water Development Board, Dhaka, Bangladesh.

SLI& NHC, 1994a. Northeast Regional Water Management Project. Fisheries Specialist Study. Volume 1: Main Report. In SNC-LAVALIN International & Northwest Hydraulic Consultants Ltd. (eds), Bangladesh Water Development Board, Dhaka, Bangladesh.

SLI& NHC, 1994b. Northeast Regional Water Management Project. Fisheries Specialist Study. Volume 2: Appendices. In SNC-LAVALIN International & Northwest Hydraulic Consultants Ltd. (eds), Bangladesh Water Development Board, Dhaka, Bangladesh.

SLI& NHC, 1994c. Northeast Regional Water Management Project. Specialist Study Agriculture in the Northeast Region. Final Report. In SNC-LAVALIN International & Northwest Hydraulic Consultants Ltd. (eds), Bangladesh Water Development Board, Dhaka, Bangladesh.

Smakhtin, V. U., 2001. Low flow hydrology: a review, Journal of Hydrology 240: 147-186.

Soussan, J., 2004a. Poverty and Water Security; Understanding How Water Affects the Poor. Asian Development Bank.

Soussan, J., 2004b. Water and Poverty; Fighting Poverty through Water Management. Asian Development Bank.

Stalnaker, C., B. L. Lamb, J. Henriksen, K. Bovee & J. Bartholow, 1995. The Instream Flow Incremental Methodology; A primer for IFIM. U.S. Department of the Interior, National Biological Service, Washington D.C.: 1-47.

Tennant, D. L., 1976. Instream flow regimens for fish, wildlife, recreation and related environmental resources, Fisheries 1: 6-10.

Tharme, R. E., 2003. A global perspective on environmental flow assessment: emerging trends in the development and application of environmental flow methodologies for rivers, River Research and Applications 19: 397-441.

Thompson, J. R.& G. Polet, 2000. Hydrology and land use in a Sahelian floodplain wetland, Wetlands 20: 639-659.

Turner, R. K., J. C. J. M. Van den Bergh, T. Soderqvist, A. Barendregt, J. Van der Straaten, E. Maltby & E. C. Van Ierland, 2000. Ecological-economic analysis of wetlands: scientific integration for management and policy, Ecological Economics 7-23.

UNDP, 2005. International Cooperation at a crossroads. Aid, trade and security in an unequal world. Human Development Report 2005. UNDP.

UNEP, 2002. Rainwater Harvesting and Utilisation. An Environmentally Sound Approach for Sustainable Urban Water Management: An Introductory Guide for Decision-Makers.

United Nations, 2003. Water for People Water for Life. UNESCO, Berghahn Books.

United Nations, 2006. Water a shared responsibility. UNESCO, Berghahn Books.

Van Beek, E.& K. S. Meijer, 2006. Integrated Water Resources Management for the Sistan Closed Inland Delta, Iran. Main Report. WL | Delft Hydraulics, Delft, The Netherlands.

Van der Hoek, W., F. Konradsen & W. A. Jehangir, 1999. Domestic use of irrigation water: health hazard or opportunity?, International Journal of Water Resources Development 15: 107-119.

Van der Maarel, E.& P. L. Dauvellier, 1978. Naar een Globaal Ecologisch Model voor de ruimtelijke ontwikkeling van Nederland (Toward an Ecological Framework for the Spatial Development of the Netherlands). *In Dutch*. Staatsuitgeverij, The Hague, The Netherlands.

Vanclay, F., 2000. Social Impact Assessment. World Commission on Dams.

Vannote, R. L., G. W. Minshall, K. W. Cummins, J. R. Sedell & C. E. Cushing, 1980. The River Coninuum Concept, Can. J. Fish. Aquat. Sci. 37: 130-137.

Vekerdy, Z.& R. Dost, 2005. History of Environmental Change in the Sistan Basin. ITC and UNEP, Geneva, Swizerland.

Wallace, J. S., M. C. Acreman & C. A. Sullivan, 2003. The sharing of water between society and ecosystems: from conflict to catchment-based co-management, Phil. Trans. Royal Society London 358: 2011-2026.

WARPO, 2000. National Water Management Plan Project. Draft Development Strategy. Ministry of Water Resources. Government of the People's Republic of Bangladesh.

WCD, 2000. Dams and Development: A New Framework for Decision-Making. The report of the World Commission on Dams: An Overview. World Commission on Dams.

References

WCED, 1987. Our Common Future. United Nations.

Whittaker, D.& B. Shelby, 2002. Evaluating Instream Flows for Recreation: Applying the Structural Norm Approach to Biophysical Conditions, Leisure Sciences 363-374.

WHO, 1999. Human Health and Dams. World Health Organization, Geneva.

WL | Delft Hydraulics, 2004. RIBASIM Version 6.32: Technical Reference Manual. WL | Delft Hydraulics, Delft, The Netherlands.

Samenvatting

Inleiding

Gedurende vele eeuwen hebben mensen de natuurlijke loop van rivieren veranderd, bijvoorbeeld door het aanleggen van stuwmeren, om hiermee het welzijn van mensen te vergroten. Tijdens de laatste eeuw is duidelijk geworden dat deze ontwikkeling van waterlichamen negatieve gevolgen heeft voor benedenstrooms van zulke ingrepen gelegen rivierecosystemen. Aangezien rivierecosystemen een bron vormen van inkomen, voedsel en andere goederen en diensten voor miljoenen mensen over de gehele wereld, hebben lokale gemeenschappen vaak te lijden gehad van deze negatieve effecten op rivierecosystemen. Hoewel geen twijfel bestaat dat de ingrepen in de rivieren voor grote groepen mensen inderdaad tot een verbetering van hun levensomstandigheden hebben geleid, waren deze voordelen vaak niet gelijk verdeeld over de diverse mensen in het stroomgebied.

De term *environmental flows* beschrijft dat deel van de natuurlijke rivierafvoer dat in de rivier, over wetlands en uiteindelijk in de kustzone stroomt, om op deze manier ecosystemen, en het gebruik van ecosystemen door mensen, in stand te houden. Diverse methoden zijn ontwikkeld om deze *environmental flows* te bepalen. De meeste van deze *environmental flow assessment* methoden richten zich op de relatie tussen de rivierafvoer en de conditie van het ecosysteem en kwantificeren niet expliciet het belang van de rivierafvoer voor het welzijn van verscheidene groepen mensen.

Om watervragen door verschillende gebruikers tegen elkaar te kunnen afwegen in het Integraal Waterbeheer is het nodig om alle positieve en negatieve effecten van het toekennen van water aan een bepaalde watergebruikssector te kwantificeren. Oorspronkelijk concentreerde deze afweging zich op economische effecten, maar tegenwoordig worden ook duurzaamheid van het milieu en sociale gelijkwaardigheid als belangrijke criteria beschouwd voor het evalueren van potentiële waterbeheersstrategieën. Om sociale gelijkwaardigheid, gedefinieerd als de verdeling van zowel positieve als negatieve effecten over diverse groepen belanghebbenden, te kunnen vaststellen, is het nodig ook het belang van *environmental flows* voor het welzijn van elk van deze groepen belanghebbenden te kwantificeren.

Methoden om de waarde van *environmental flows* voor menselijk welzijn (verder menselijke welzijnswaarden van *environmental flows* genoemd) te kwantificeren zijn momenteel niet beschikbaar, maar zijn hard nodig om gelijkwaardig waterbeheer te faciliteren en te versterken.

Samenvatting

De onderzoeksdoelstelling voor dit proefschrift was daarom:

> *Het ontwikkelen van een aanpak om de menselijke welzijnswaarden en sociale gelijkwaardigheid gerelateerd aan environmental flows te kunnen vaststellen voor toepassing in het Integraal Waterbeheer.*

Om deze doelstelling te bereiken is een conceptueel model ontwikkeld dat de relaties tussen menselijk welzijn, rivierecosystemen en rivierafvoeren beschrijft. Dit model is vervolgens toegepast in twee case studies, de eerste over de rivieruiterwaarden van de Surma en de Kushiyara rivieren in Bangladesh en de tweede over de Hamoun wetlands in Iran. Beide case studies hadden tot doel het conceptuele model the testen en te verfijnen.

Conceptueel model

Om menselijk welzijn te beschrijven zijn verscheidene onderdelen van dit concept onderscheiden. Drie onderdelen zijn rechtstreeks gerelateerd aan het rivierecosysteem: 1) inkomen & voedsel, 2) gezondheid en 3) beleving. Omdat het socio-economische systeem complex en dynamisch is, is het waarschijnlijk dat veranderingen in deze drie onderdelen van menselijk welzijn effecten zullen hebben op andere onderdelen van menselijk welzijn zoals onafhankelijkheid, sociale structuur en andere psycho-sociale factoren. Deze onderdelen van welzijn worden in dit proefschrift twee orde waarden genoemd, omdat ze alleen indirect beïnvloed worden door veranderingen in het rivierecosysteem. De onderdelen die direct gerelateerd zijn aan het ecosysteem worden eerste orde waarden genoemd. Veranderingen in tweede orde onderdelen kunnen resulteren in verdere veranderingen in de eerste orde onderdelen, zowel in een positieve als in een negatieve richting. De kwantificering in dit proefschrift is gericht op de eerste orde waarden.

Het kwantificeren van effecten van veranderingen in het rivierafvoerregime op menselijk welzijn vraagt een diepgaand conceptueel begrip van de *relaties* tussen rivierafvoeren, het rivierecosysteem en menselijk welzijn. Hoewel dit proefschrift vooral ingaat op de relatie tussen menselijk welzijn en het rivierecosysteem, was het nodig om ook de relatie tussen het rivierecosysteem en het rivierafvoerregime in het conceptuele model op te nemen. Beide relaties zijn nodig om veranderingen in menselijk welzijn als gevolg van veranderingen in het afvoerregime te kunnen kwantificeren. Bovendien, alleen door alle relaties te kwantificeren was het mogelijk om problemen in de aansluiting tussen verschillende concepten en methoden te identificeren.

Om het belang van het rivierafvoerregime voor het welzijn van diverse groepen mensen te kunnen vaststellen, is het nodig niet alleen aandacht te besteden aan de relaties tussen welzijn, rivierecosystemen en het rivierafvoerregime, maar ook aan de context van deze relaties. De context omvat alle factoren die deze relaties kunnen beïnvloeden. Een voorbeeld van de context is toegang tot ecosysteemgoederen en -diensten. Wanneer arme vissers geen toegang hebben tot bepaalde visgebieden, zal deze specifieke groep vissers niet profiteren van het in stand houden van dit deel van het ecosysteem. Alleen door voldoende aandacht aan de context te besteden kan het belang van het rivierecosysteem, en van rivierafvoeren, voor menselijk welzijn worden begrepen.

Zowel de relaties als de context zijn opgenomen in het conceptuele model dat in dit proefschrift ontwikkeld is (zie Figuur S.1).

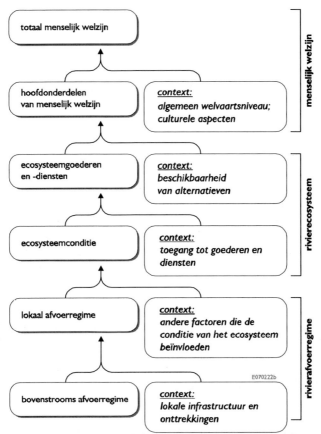

Figuur S.1 Het conceptuele model met de relaties tussen menselijk welzijn, het rivierecosysteem en het rivierafvoerregime, en de context van deze relaties.

Het doel van het conceptuele model is het ondersteunen van de bepaling van effecten op menselijk welzijn en van de sociale gelijkwaardigheid van alternatieve waterbeheersstrategieën. Het is belangrijk om de relaties van het conceptuele model te kwantificeren voor de diverse groepen mensen waarvoor verwacht kan worden dat veranderingen in het rivierafvoerregime voor hen tot verschillende effecten zullen leiden. Het identificeren van deze groepen mensen (de stakeholder groepen) is daarom een belangrijke eerste stap in de toepassing van het conceptuele model.

De toepassing van het conceptuele model bestaat in totaal uit vijf stappen:

1. het identificeren van stakeholder groepen;
2. het vaststellen van de relatie tussen menselijk welzijn en het rivierecosysteem;
3. het vaststellen van de relatie tussen het rivierecosysteem en het lokale afvoerregime;

4. het vaststellen van de relatie tussen het lokale afvoerregime en het bovenstroomse afvoerregime; en
5. het bepalen van de effecten van waterbeheersmaatregelen op het welzijn van de geïdentificeerde stakeholder groepen.

Case studies

Surma and Kushiyara Rivieren en uiterwaarden, Bangladesh

In de Surma en Kushiyara Rivieren en uiterwaarden in Noordoost Bangladesh wordt verwacht dat het afvoerregime zal wijzigen als gevolg van drie ontwikkelingen:

- Morfologische veranderingen op de locatie waar de Barak Rivier Bangladesh binnenstroomt vanuit India en splitst in de Surma Rivier en de Kushiyara Rivier.
- De constructie van het Tipaimukh stuwmeer in de Barak Rivier (bovenstrooms van de grens tussen India en Bangladesh). Verwacht wordt dat dit leidt tot gereduceerde piekafvoeren en verhoogde laagwaterafvoeren.
- De onttrekking van water voor irrigatie van de Cachar Vlakte vanuit de Barak Rivier (eveneens bovenstrooms van de grens tussen India en Bangladesh). In combinatie met het Tipaimukh stuwmeer wordt verwacht dat deze ontwikkeling zal resulteren in een verdere afname van piekafvoeren. De toename in laagwaterafvoeren als gevolg van het Tipaimukh stuwmeer zal verkleind worden. Het is niet duidelijk of het netto effect op de laagwaterafvoeren een toe- of een afname zal zijn.

Het conceptuele model is toegepast om het belang van het rivierafvoerregime voor de ongeveer 300.000 mensen die in de Surma-Kushiyara uiterwaarden wonen vast te stellen. Gegevensverzameling voor de relatie tussen menselijk welzijn en het ecosysteem is gedaan door interviews op het niveau van huishoudens. Voor de andere relaties zijn bestaande gegevens gebruikt.

Drie belangrijke stakeholder groepen kunnen worden onderscheiden in de Surma-Kushiyara uiterwaarden: boeren, vissers en anderen. Boeren en vissers zijn voor hun inkomen en voedselvoorziening grotendeels afhankelijk van de goederen en diensten van de rivier en de uiterwaarden. Ongeveer 10% van de bewoners gebruikt de rivier voor huishoudelijke doeleinden. Dit zijn vaak de armste mensen: de vissers en de landloze arbeiders. De belangrijkste relatie tussen het rivierecosysteem en de beleving zijn de ongemakken die mensen ondervinden als gevolg van overstroming van de bebouwde gebieden. Ook voor dit welzijnsonderdeel ondervinden de armste mensen de grootste effecten.

In een kwalitatieve analyse zijn de verwachte effecten van de drie ontwikkelingen die tot veranderingen in het afvoerregime leiden onderzocht. De resultaten laten zien dat vooral vissers inkomen zullen kwijtraken wanneer de uiterwaarden niet op tijd overstromen. Aan de andere kant zal de gezondheid van de vissers en van andere arme huishoudens profiteren van een toename van de rivierafvoer tijdens het droge seizoen.

Hamoun wetlands, Iran

Een tweede toepassing van het conceptuele model is uitgevoerd als onderdeel van een Integraal Waterbeheerstudie in de Sistan delta in Iran. Deze delta is omgeven door het Hamoun wetlands systeem, dat een oppervlak kan hebben van 500.000 ha onder natte omstandigheden. De wetlands zijn van grote waarde voor zowel de mensen die de wetlands voor inkomengeneratie gebruiken, als voor andere mensen die profiteren van het voorkomen van zandstormen en van de achtergrond voor festiviteiten zoals bijvoorbeeld Nieuwjaarsvieringen.

De vijf stappen van het conceptuele model zijn gevolgd om de welzijnswaarden te berekenen die het gevolg zijn van diverse waterbeheersstrategieën. Groepdiscussies en een enquête zijn uitgevoerd om gegevens te verzamelen over de relatie tussen menselijk welzijn en het wetlandecosysteem. Voor de andere relaties is gebruik gemaakt van de resultaten van andere onderdelen van de Integraal Waterbeheerstudie.

Met betrekking tot hun afhankelijkheid van de Hamoun wetlands kan de bevolking van de Sistan delta worden onderverdeeld in drie groepen. Van de totaal ongeveer 71.000 huishoudens, zijn ongeveer 15.000 huishoudens rechtstreeks afhankelijk van de wetlands voor meer dan 70% van hun inkomen, ongeveer 30.000 huishoudens zijn voor meer dan 70% van hun inkomen afhankelijk van landbouw, terwijl het inkomen van de overgebleven 26.000 stedelijke huishoudens grotendeels helemaal niet van water afhankelijk is.

De belangrijkste ecosysteemgoederen en -diensten die door de bevolking gebruikt worden zijn de beschikbaarheid van vis, vogels en riet voor inkomen en voedselproductie, het voorkomen van zandstormen en het reguleren van het lokale klimaat. Vijf hydrologische parameters zijn geïdentificeerd die van belang zijn om deze goederen en diensten in stand te houden. Voor het beschermen van vis, riet en vogels zijn dit het overlopen van het wetland in een benedenstroomse rivier, regelmatige droogtes, een minimaal overstroomd oppervlak in het najaar en een minimaal instromend volume aan water in het voorjaar. Daarnaast wordt het van belang geacht om ieder jaar tussen mei en augustus een minimaal oppervlak van de Hamoun-e-Saberi, één van de wetlands, overstroomd te hebben om hiermee het ontstaan van zandstormen te voorkomen.

Waterbeheersstrategieën kunnen het welzijn van de geïdentificeerde stakeholder groepen beïnvloeden door hun gevolgen voor het hydrologische regime van het wetland, de daaruitvolgende veranderingen in het ecosysteem en de resulterende beschikbaarheid van ecosysteemgoederen en -diensten. Voor alle onderdelen van menselijk welzijn bespreekt dit proefschrift een benadering voor het kwantificeren van de effecten hierop. Met deze relaties zijn de effecten van alternatieve waterbeheersstrategieën op het welzijn van de diverse stakeholder groepen en de gelijkwaardigheid in verdeling van deze effecten bepaald.

De omvangrijkste watervragen in het gebied komen van de geïrrigeerde landbouw en de Hamoun wetlands. Beide watergebruikssectoren zijn belangrijk voor de economie van de regio en het welzijn van de stakeholder groepen. Een balans moet gevonden worden voor de verdeling van water over deze twee sectoren. Om deze balans te onderzoeken zijn de effecten van een vergroting en een verkleining van het geïrrigeerde gebied gekwantificeerd. De resultaten laten zien dat in de huidige situatie de verdeling van effecten gelijkwaardiger is dan wanneer de twee strategieën worden toegepast. Hoewel in de huidige situatie en in de situatie met een kleiner oppervlak geïrrigeerde landbouw de berekende conditie van het ecosysteem gelijk is, varieert het aantal jaren dat de verscheidene stakeholdergroepen te lijden hebben onder een gereduceerd inkomen. Dit geeft aan dat het bepalen van effecten op menselijk welzijn voorziet in nuttige aanvullende informatie die geïnformeerd en gelijkwaardig waterbeheer versterkt.

Conclusies

Door het beschrijven van zowel de relaties tussen menselijk welzijn, het rivierecosysteem en het rivierafvoerregime, als de context, zal het voorgestelde conceptuele model de kwantificering van veranderingen in het rivierafvoerregime of het rivierecosysteem in termen van menselijk welzijn en sociale gelijkwaardigheid in een Integraal Waterbeheeranalyse vereenvoudigen en structuren.

Hoewel het conceptuele model en de stapsgewijze aanpak generiek toepasbaar zijn, zullen de methoden voor gegevensverzameling en analyse afgestemd moeten worden op de situatie. In het droge Sistan was het makkelijk om de grenzen van het wetlandecosysteem te definiëren en mensen te identificeren die van dit wetland gebruik maakten. In het natte noord-oosten van Bangladesh was het moeilijker om te identificeren in welke mate de gebruikte ecosysteemgoederen en -diensten van het rivierafvoerregime afhingen, en daarmee of de gebruikers van de goederen en diensten stakeholders waren met betrekking tot het afvoerregime. Bovendien kon vanwege het feit dat in Sistan veranderingen al hadden plaatsgevonden direct de discussie worden aangegaan met de stakeholders over de diverse effecten op hun welzijn als gevolg van de veranderingen in het wetland. In de Surma-Kushiyara uiterwaarden, waar veranderingen nog niet hadden plaatsgevonden, richtte de gegevensverzameling zich op het huidige gebruik van het ecosysteem en het begrijpen van de context om effecten op welzijn als gevolg van veranderingen in het afvoerregime te voorspellen.

Uit de case studies is gebleken dat door het bepalen van menselijke welzijnswaarden van het rivierecosysteem nuttige aanvullende informatie is gegenereerd om besluitvorming te ondersteunen. Aanbevolen is daarom om dergelijke analyses altijd uit te voeren als onderdeel van een Integraal Waterbeheerstudie. Wanneer tijd en andere middelen beperkt beschikbaar zijn, kan de analyse van het menselijk welzijn worden uitgevoerd als enkel een kwalitatieve bepaling met vooral aandacht voor het identificeren van stakeholder groepen en het belang van de diverse ecosysteemgoederen en -diensten voor elk van deze groepen. Wanneer tijd en overige middelen in voldoende mate beschikbaar zijn kan de kwantificering van eerste orde effecten, zoals beschreven in dit proefschrift, worden uitgebreid met een kwantificering van tweede orde effecten.

De algehele conclusie is dat dit proefschrift voorziet in een aanpak voor het bepalen van menselijke welzijnswaarden samen met een stapsgewijze benadering voor de assessment zelf. Met deze aanpak draagt dit proefschrift bij aan de verdere operationalisatie van de concepten van Integraal Waterbeheer en *environmental flows*. De in dit proefschrift gepresenteerde aanpak genereert uitgebreide en sociaal-relevante informatie voor besluitvormers, wat essentieel is om sociale gelijkwaardigheid in het waterbeheer te versterken.

Bengali summary
Translated by: Asha Naznin

বাংলা সারসংক্ষেপ
অনুবাদ: আশা নাজনীন

সূচনা

কয়েক শতাব্দী ধরেই জলবন্টন ব্যবস্থার উন্নয়ন মানুষের জীবনধারণের সুব্যবস্থার অন্যতম ধারক হিসেবে বিবেচিত হয়ে আসছে। নদী ব্যবস্থার উন্নয়ন তথা জলসংস্থানের উর্ধ্বস্রোত বাধাগ্রস্থ হবার ফলস্বরুপ বাস্তসংস্থানে যে নেতিবাচক প্রভাব পড়েছে, তা গত শতাব্দীতেই স্পষ্টভাবে প্রতীয়মান হয়। সারা বিশ্ব জুড়ে কয়েক নিযুত মানুষের জীবিকা, অন্ন এবং সম্পদের একটি অন্যতম উৎস জলীয় বাস্তসংস্থান। যদিও জলসংস্থান ব্যবস্থায় উন্নয়ন নি:সন্দেহে ভাবে একটি বিরাট সংখ্যক মানুষের জীবনের উন্নত ব্যবহারের ধারক হিসেবে সূচিত হয়েছে, তারপরেও নদী তীরবর্তী সকল মানুষের মধ্যে এর সুষ্ঠু বন্টন এখনো হয়নি। যেহেতু নদীতীরবর্তী মানুষের জীবিকা অনেকাংশেই নদীর ওপর নির্ভরশীল, ফলে জলবন্টন ব্যবস্থার বিভিন্ন বাঁধা বিপত্তিগুলোর ক্ষতিকর প্রভাব তাদের ওপরেই পড়েছে।

পরিবেশের স্বাভাবিক প্রবাহ, গতির নির্ভরতা অনেকাংশেই নদীর পানি এবং তীরবর্তী ভূমির প্রতি আলোকপাত করে, যেগুলো বাস্তসংস্থানের অন্যতম উপাদান হিসেবে বিবেচিত। অনেকাংশেই এই উপাদানসমূহ মানুষের সার্বিক সুযোগ সুবিধার সংস্থান করে থাকে। পরিবেশের স্বাভাবিক প্রবাহ ধারা বজায় রাখার নিমিত্তে প্রয়োজনীয় উপাদান সমূহের পরিগণনের অনেক উপায়ই বিভিন্ন সময় সূচিত হয়েছে। মানুষের উন্নত জীবন ধারার গুরুত্বকে বিবেচনা না করে এই প্রক্রিয়াগুলোর সিংহভাগই জোর আরোপ করে নদীর নাব্যতা এবং বাস্তসংস্থানের সহসম্পর্কের ওপরে।

ইন্টিগ্রেটেড ওয়াটার রিসোর্স ম্যানেজমেন্ট (আইডব্লিউআরএম) এ পানি বন্টনের ভারসাম্য সাধনের জন্য সকল সুবিধাসমূহ এবং কোনো নির্দিষ্ট সেক্টরে পানি বন্টনের নেতিবাচক প্রভাবটিও পরিগণন করা জরুরী। প্রথাগতভাবে নদীর স্বাভাবিক প্রবাহ ব্যবস্থার কেন্দ্রবিন্দু ছিলো অর্থনৈতিক সুবিধা, কিন্তু বর্তমান সময়ে পরিবেশের স্থিতিবস্থা এবং সামাজিক ক্ষমতায়ন জলবন্টন ব্যবস্থার কৌশল সমূহের মূল্যায়নসূচক হিসেবে উপস্থিত হয়েছে। সামাজিক সাম্যসাধন তথা বিভিন্ন পুঁজিজীবী দলের প্রতি বন্টনের ইতিবাচক ও নেতিবাচক প্রভাবের পরিলক্ষণ সম্ভব করার জন্য এইসব পুঁজিজীবী দলের জীবনধারণের সুব্যবস্থার নিমিত্তে পরিবেশের স্বাভাবিক স্রোতের গুরুত্বটাও পরিগণন আবশ্যক।

Bengali summary

উন্নত জীবনযাপনের জন্য পরিবেশের প্রবাহ গতির মূল্যায়ন প্রক্রিয়া (কিংবা বলা যায় পরিবেশের প্রবাহগতির উন্নতজীবনধারক মান নির্ণয়) বর্তমানে প্রাপ্তি সাধ্য নয়, এর জন্য এখনো প্রয়োজন নিরপেক্ষ জলবন্টন ব্যবস্থার সুবিধা প্রণয়ন এবং বিস্তরণ।

এই তত্ত্বের পরীক্ষণ উদ্দেশ্য ছিলো:

ইনটিগ্রেটেড ওয়াটার রিসোর্স ম্যানেজমেন্ট (আইডবিউআরএম) এর প্রয়োগের জন্য জীবনধারনের সুব্যবস্থা এবং সামাজিক ক্ষমতায়নের মূল্যায়নের নিমিতে পদক্ষেপের উন্নয়ন যা পরিবেশের প্রবাহগতির সঙ্গে সম্পর্কিত।

এই উদ্দেশ্য সাধনের জন্য পরিবেশের প্রবাহগতি, মানবজীবনের সুব্যবস্থা এবং আইডবিউআরএমের সম্পর্কে একটি কাল্পনিক/ধারণাগত নকশা প্রণীত হয়। অত:পর, দুটো কেস স্টাডির মাধ্যমে তা পরীক্ষীত হয়। এর প্রথমটি ছিলো বাংলাদেশে সুরমা কুশিয়ারা নদীর পাবন সমভূমি অঞ্চল এবং ইরানের হামৌন উপকূলীয় সমভূমিয় অঞ্চল।

কাল্পনিক নকশা

উন্নত জীবন ব্যবস্থাকে সংজ্ঞায়িত করার জন্য এর বিভিন্ন উপাদানকে চিহ্নিত করা হয়। তিনটি উপাদান নদীর বাস্তুসংস্থানের সঙ্গে সরাসরি সম্পৃক্ত। (১) অন্ন ও জীবিকা (২) স্বাস্থ্য এবং (৩) প্রত্যক্ষণ এবং অভিজ্ঞতা। যেহেতু আর্থসামাজিক ব্যবস্থা একটি জটিল এবং গতিশীল প্রক্রিয়া, উন্নত জীবনধারার একটি উপাদানের পরিবর্তন, সেহেতু এটি অন্যান্য আরো উপাদানের ওপর প্রভাব বিস্তার করে। স্বাধীনতা, সামাজিক অবকাঠামো এবং অন্যান্য আরো মনো-সামাজিক বিষয়ের মতো এই উপাদানগুলো দ্বিতীয় মাত্রার উপাদান বা মানদন্ড হিসেবে বিবেচ্য। কারণ এরা নদী বাস্তু সংস্থানে সরাসরি প্রভাবিত করেনা। বাস্তুসংস্থানের সঙ্গে সরাসরি সম্পৃক্ত উপাদানগুলো তাই প্রথম মাত্রার উপাদান হিসেবে চিহ্নিত। দ্বিতীয়মাত্রার উপাদানের পরিবর্তনের ফলশ্রুতিতে প্রথম মাত্রার উপাদানগুলোর পরিবর্তন হতে পারে- ধনাত্মক কিংবা ঋণাত্মক। এই তত্ত্বে সংখ্যাতাত্ত্বিক গণনা আলোকপাত করে প্রথম মাত্রার উপাদানের ওপরেই।

উন্নত জীবনধারার পরিবর্তনের ওপর প্রবাহমান ধারার প্রভাবের পরিমান নির্ধারণ করতে গেলে নদীপ্রবাহ, বাস্তুসংস্থান এবং উন্নত জীবনধারার মধ্যে বিদ্যমান সম্পর্ক সম্পর্কে সম্যক ধারণা এবং অনুধাবন অতীব প্রয়োজন। যদিও এই তত্ত্বটি মানবজীবন ধারা এবং বাস্তুসংস্থানের মধ্যে সম্পর্কের প্রতি জোর দিয়েছে। তার সঙ্গে বাস্তুসংস্থান এবং নদী প্রবাহ ধারার ব্যবস্থাপনায়, কাল্পনিক নকশার সম্পর্কটিও প্রণীত হওয়া জরুরী। দুটো সম্পর্কই প্রবাহ ধারার ফলস্বরূপ মানবজীবনধারায় সংঘটিত পরিবর্তনকে সূচিত করে। কেবল সম্যক সম্পর্ক সূচনা করার মাধ্যমেই কল্পনা এবং বাস্তবায়নের মধ্যে অসামঞ্জস্যতাকে চিহ্নিত করা সম্ভব।

জনগণের উন্নত জীবনব্যবস্থার লক্ষ্যে নদীপ্রবাহ ব্যবস্থাপনা এবং নদীতীরবর্তী বাস্তুসংস্থানের গুরুত্বকে মূল্যায়ন করার জন্য কেবল উন্নত জীবন ব্যবস্থা, বাস্তুসংস্থান এবং নদী ব্যবস্থার মধ্যে সম্পর্ক নির্ণয় করলেই চলবেনা। সেই সঙ্গে সম্পর্কের সূচকগুলোকেও সংযুক্ত করতে হবে। পারস্পরিক সম্পর্কগুলোর

সূচকসমূহের দিকে মনোযোগ নিক্ষেপণের মাধ্যমেই উন্নত জীবনধারার সঙ্গে নদীর স্বাভাবিক প্রবাহধারা ও বাস্তুসংস্থানের সম্পর্কের গুরুত্ব অনুধাবন সম্ভব।

কাল্পনিক নকশার কাজ হলো উন্নত জীবনব্যবস্থার মানদন্ড এবং সামাজিক ক্ষমতায়নের ওপর বিকল্প জলবন্টন ব্যবস্থাপনা কৌশলের প্রভাবের বিশ্লেষণকে সমর্থন করা। কাল্পনিক নকশার বিভিন্ন সূত্রের মধ্যে সংযোগ ব্যবস্থাপনা এই জন্য জরুরী যে, বিভিন্ন ধরনের মানুষ এই পরিবর্তনের বিভিন্ন প্রভাব অভিজ্ঞান করতে যাচ্ছে। এই বিভিন্ন পুঁজিজীবী দলগুলোকে চিহ্নিত করাটা কাল্পনিক নকশা বাস্তবায়নেরই একটি গুরুত্বপূর্ণ পদক্ষেপ।

কাল্পনিক নকশার বাস্তবায়ন প্রক্রিয়া ৫টি পদক্ষেপের সমন্বয়ে গঠিত-

১. পুঁজিজীবী দলগুলোকে চিহ্নিতকরণ,
২. উন্নতজীবনধারা এবং নদীর বাস্তুসংস্থানের মধ্যে সম্পর্ক নির্ণয়
৩. নদীর বাস্তুসংস্থান এবং স্থানীয় প্রবাহ ব্যবস্থা (জলীয় বৈশিষ্ট্যসমূহ) এর মধ্যে সম্পর্ক নির্ণয়।
৪. স্থানীয় প্রবাহ ব্যবস্থা এবং ঊর্ধ্বগামী প্রবাহ ব্যবস্থার মধ্যে বিদ্যমান সম্পর্কের মূল্যায়ন
৫. চিহ্নিত পুঁজিজীবী দলের সুষ্ঠু জীবনধারার ওপর জলবন্টন ব্যবস্থার পরিবর্তনের প্রভাব অনুমান

কেস স্টাডিস

সুরমা ও কুশিয়ারা নদী, বাংলাদেশ

তিনটি উন্নয়ন কার্যক্রমের ফলস্বরূপ বাংলাদেশের উত্তর পূর্বাঞ্চলীয় সুরমা ও কুশিয়ারা নদীর স্রোতধারায় পরিবর্তন আসবে বলে আশা করা যায়

- যে স্থানে বারাক নদী ভারত থেকে বাংলাদেশে প্রবেশ করে এবং সুরমা ও কুশিয়ারা নদীকে দ্বিবিভক্ত করে, তার আঙ্গিক পরিবর্তন
- বারাক নদীতে টিপাঈমুখ বাঁধের স্থাপন ঊর্ধ্বস্রোতকে হ্রাস করে অর্ধস্রোতকে বৃদ্ধি করবে বলে ধারণা করা হয়।
- টিপাঈমুখ জলাশয়ের সঙ্গে সঙ্গতি রেখে বারাক নদী থেকে কাছর সমভূমি পর্যন্ত জলনিষ্কাশন ব্যবহারের ভিন্নমুখীকরণ থেকে ধারণা করা হয় যে, ফলশ্রুতিতে তা ঊর্ধ্বস্রোতকে আরো হ্রাস করবে। টিপাঈমুখ জলাধার থেকে উৎসারিত নিম্নস্রোতও সীমিত হবে। তবে এটা এখনো অনিশ্চিত যে, অর্ধস্রোতের ওপর সার্বিক প্রভাব বৃদ্ধিকর হবেনা, হ্রাসকর হবে।

সুরমা কুশিয়ারা তীরবর্তী পাবন সমভূমির প্রায় ৩০০,০০০ (৩ লাখ) মানুষের জন্য জলবন্টন ব্যবস্থার মূল্যায়নের কাল্পনিক নকশাটি প্রয়োগ করা হয়। সাক্ষাৎকার নেওয়ার মাধ্যমে উন্নত জীবব্যবস্থা ও বাস্তুসংস্থানের সংযোগ সম্পর্কিত উপাত্ত সংগ্রহ করা হয়। অন্যান্য সংযোগের জন্য আপাত বিদ্যমান উপাত্ত সমূহ ব্যবহার করা হয়।

প্রধানত তিন ধরনের পুঁজিজীবী দল চিহ্নিত করা হয় সুরমা কুশিয়ারা তীরবর্তী সমভূমিতে-কৃষক, জেলে এবং অন্যান্য।

কৃষক এবং মৎস্যজীবী গোষ্ঠী অনেকাংশেই নদী এবং নদীতীরবর্তী সমভূমি থেকে প্রাপ্ত সম্পদের ওপর নির্ভর করে জীবিকা সংস্থান করে। অধিবাসীদের প্রায় ১০ শতাংশ তাদের গৃহব্যবস্থাপনার জন্য নদীর পানি ব্যবহার করে। গৃহপালনের জন্য নদীর ওপর নির্ভরশীল এই মানুষগুলো একেবারেই হতদরিদ্র যাদের মধ্যে রয়েছে মৎস্যজীবী এবং ভূমিহীন শ্রমিক। নদীর বাস্তুসংস্থান এবং প্রত্যক্ষণ ও অভিজ্ঞতার মধ্যে সবচেয়ে বড় সংযোগ হলো এই অঞ্চলের প্লাবিত হবার সম্ভাবনার সঙ্গে সম্পর্ক এবং এই জীবনধারক মানদন্ডটির কারণে সবচেয়ে দরিদ্র লোকগুলোই সবচেয়ে বেশি প্রভাবিত হয়।

একটি আন্তিক বিশ্লেষনের সময় প্রবাহ ব্যবস্থার উপরিলিখিত ৩টি পরিবর্তনের কথাও বিবেচনা করা হয়। এই বিশ্লেষনটি এটা প্রকাশ করে যে, প্লাবন সমভূমিতে সঠিক সময়ে বন্যা না হলে সবচেয়ে বেশি ক্ষতিগ্রস্ত হবে মৎস্যজীবী গোষ্ঠী। অপরদিকে অন্যান্য দরিদ্র অধিবাসীদের স্বাস্থ্যগত সুবিধা প্রণীত হবে যদি সুষ্ক ঋতুতে নদীর জলপ্রবাহ বৃদ্ধি করা যায়।

হামৌন উপকূলীয় অঞ্চল, ইরান

আইডবিউআরএমের গবেষণার অংশ হিসেবে প্রণীত কাল্পনিক নকশার দ্বিতীয় প্রয়োগটি ছিলো ইরানের সিস্তান বদ্বীপ অঞ্চলে। ৫ লক্ষ হেক্টর আয়তনের এই বদ্বীপ অঞ্চলটি হামৌন উপকূলীয় সমভূমি দ্বারা বেষ্ঠিত। এর উপকূলীয় সমভূমি দুই ধরনের মানুষের জন্য সমানভাবে মূল্যবান। এক- যারা জীবিকা সংস্থানের জন্য এই অঞ্চলটাকে ব্যবহার করে, দুই- যারা এর বালিভূমি সংরক্ষণের মাধ্যমে এখানে বাণিজ্য স্থাপনের সুযোগকে কাজে লাগায়।

কাল্পনিক নকশার ৫টি পদক্ষেপ অনুসরণের মাধ্যমে বিভিন্ন জলবন্টন ব্যবহার কৌশলের অধীনে উন্নত জীবনধারকের একটি পরিমাণগত মূল্যায়ন নির্ধারণের চেষ্টা করা হয়। দলগত আলোচনা এবং জরীপের মাধ্যমে উন্নত জীবনধারা এবং সমভূমি বাস্তুসংস্থানের সংযোগ সম্পর্কিত উপাত্ত সংগ্রহ করা হয়। অন্যান্য সংযোগের জন্য আইডবিউআরএম এর গবেষণার বিভিন্ন বিশ্লেষনাত্মক উপাত্ত প্রয়োগ করা হয়।

হামৌন সমভূমির ওপর নির্ভরতাকে বিবেচনা করলে সিস্তান বদ্বীপের জনগণকে ৩ ভাগে ভাগ করা যায়। প্রায় ৭১ হাজার অধিবাসীর মধ্যে মোটামুটি ১৫ হাজার অধিবাসী সরাসরিভাবে তাদের জীবিকা সংস্থানের জন্য হামৌন উপকূলীয় মরুভূমির ওপরে ৭০শতাংশের বেশি নির্ভর করে। মোটামুটি ৩০ হাজার মানুষ তাদের ৭০ শতাংশের বেশি জীবিকার জন্য নির্ভর করে নিষ্কাশিত কৃষি ব্যবহার ওপরে। অপরদিকে বাকি ২৬ হাজার শহুরে অধিবাসী তাদের আয়ের জন্য নদীর ওপর খুব কমই নির্ভর করে।

এই অঞ্চলের জনগণের সম্পদ এবং জীবিকার প্রধান উৎস হচ্ছে মাছ, পাখি এবং নলখাগড়ার পর্যাপ্ততা যা তাদের আয় এবং খাদ্য উপাদানকে সমর্থন দান করে। সেই সঙ্গে বেলাভূমির সংরক্ষণ এবং স্থানীয় জলবায়ুর স্থিতিশীলতা ও অনেক বড় ভূমিকা পালন করে থাকে। ৫টি গুরুত্বপূর্ণ জলীয় মানদন্ড নির্ধারণ করা হয় যা সম্পদ এবং জীবিকার স্থিতিশীলতাকে বাধাগ্রস্ত করে। এগুলো হচ্ছে সমভূমির ভাঙন, নিয়মিত অনাবৃষ্টি, শরতে একেবারেই কম প্লাবন এবং বসন্তে নদী প্রবাহে নিম্নগতি। বালিভূমি সংরক্ষণের জন্য হামৌন-ই- সাবেরি অঞ্চলে কিছু অংশ হলেও মে থেকে আগষ্ট মাসে প্লাবিত করা প্রয়োজন বলে বিবেচিত করা হয়।

আর্দ্র উপকূলীয় অঞ্চলে জলবন্টন ব্যবস্থা বাস্তসংস্থানে প্রতিনিয়ত পরিবর্তন এবং এর ফলশ্রুতিতে বাস্তুসংস্থানের বিভিন্ন উপাদানের পর্যাপ্ততার অভাবের মাধ্যমে পানিবন্টন ব্যবস্থার কৌশলসমূহ চিহ্নিত পুজিজীবী দলের উন্নত জীবনধারাকে প্রভাবিত করতে পারে। উন্নতজীবনধারার সকল উপাদানকে বিবেচনায় রেখে বাস্তুসংস্থানের পরিবর্তনের ফলস্বরূপ উন্নত জীবনধারায় কি পরিবর্তন সাধিত হয় তা পরিগণনের পদক্ষেপ সম্পর্কে আলোচনা করা হয়। সেই সূত্রে পুজিজীবী দলের ওপর তার প্রভাব এবং সেই প্রভাবের সমবন্টনের মূল্যায়নও করা হয়।

নিষ্কাশিত কৃষি ব্যবস্থার ক্ষেত্র এবং হামেন উপকূলীয় অঞ্চলে পানির সবচেয়ে বেশি চাহিদা দেখা যায়। ওই অঞ্চলের অর্থনীতিতে এই দুই গ্রাহকই সমান গুরুত্বপূর্ণ ভূমিকা পালন করে, যা পুজিজীবী গোষ্ঠীগুলোর উন্নত জীবনধারাকেও সূচিত করে। দুই দলের চাহিদানুযায়ী বন্টনের ক্ষেত্রে ভারসাম্য বজায় রাখার ব্যাপারটি এখানে জোর দেওয়া হয়। বিশেষিত কৌশলসমূহ থেকে নিষ্কাশিত অঞ্চলের হ্রাসবৃদ্ধির মাধ্যমে ভারসাম্য প্রণয়ন করার বিষয়টি আলোকপাত করা হয়। দেখা যায় যে, বর্তমান পরিস্থিতিতেই বন্টনের প্রভাব সবচেয়ে সমানভাবে করা যায়। যদি বর্তমান পরিস্থিতি এবং হ্রাসকৃত নিষ্কাশিত পরিস্থিতি বাস্তুসংস্থানের ক্ষেত্রে অপরিবর্তিত থাকবে বলে হিসাব করা হয়, বিভিন্ন পুজিজীবীগোষ্ঠীর হ্রাসকৃত আয় বিভিন্ন বছরে বিভিন্ন হবে বলেই প্রতীয়মান হয়। অর্থাৎ এটা বলা যায় যে, উন্নত জীবনধারার বিভিন্ন প্রভাবসমূহ মূল্যায়নের মাধ্যমে উপকারি তথ্য পাওয়া যায় যা তথ্যসমৃদ্ধ এবং ক্ষমতাবিধায়ক ব্যবস্থাপনার সম্ভাবনাকে বৃদ্ধি করে।

উপসংহার

মানুষের জীবনধারণের সুব্যবস্থার সঙ্গে নদীর বাস্তুসংস্থান এবং নদী প্রবাহ অঞ্চলের সম্পর্কের ভেতর দিয়ে প্রস্তাবিত কাল্পনিক নকশাটি নদীপ্রবাহ অঞ্চল অথবা নদীর বাস্তুসংস্থানের সংখ্যাতাত্ত্বিক পরিবর্তনের সুবিধা তুলে ধরবে মানুষের সুষ্ঠু জীবনব্যবস্থা এবং সামাজিক সমতার মাধ্যমে, যা হবে আইডব্লিউআরএম এর বিশ্লেষনের একটি অংশ। ধারণাগত নকশা এবং ধাপে ধাপে অগ্রসরমান পদ্ধতি সাধারণভাবে ব্যবহারের পাশাপাশি তথ্য সংগ্রহ এবং বিশ্লেষণ অবশ্যই প্রয়োজনীয়। সিস্টান মরুভূমিতে জলজ বাস্তুসংস্থান এবং মানুষ যারা জলাভূমির ওপর নির্ভরশীল তাদের চিহ্নিত করা সম্ভব। বাংলাদেশের উত্তরাঞ্চলের জলাভূমিতে দেখা গেছে, নদীপ্রবাহে বাস্তুসংস্থানের উপাদান এবং কর্মপদ্ধতিকে চিহ্নিত করা খুব কষ্টকর। ফলশ্রুতিতে, এটা খুঁজে পেতে আরো বেশি কষ্টকর যে, পুজিজীবীরা কি এই উপাদান এবং কর্মপদ্ধতি ব্যবহারকারী কি না। উপরন্তু, এটাও এখন একটা বিষয় যে, সিস্টানে ইতিমধ্যে নদীর গতিধারা মানুষের জীবনের সঙ্গে সম্পৃক্ত হয়ে গেছে, সুষ্ঠু জীবন যাপনের লক্ষ্যে উপকূলীয় পরিবর্তনের নানা প্রভাব পুঁজিজীবীদের সঙ্গে সরাসরি আলোচনার জন্য তাৎপর্যপূর্ণ হয়ে উঠেছে। সুরমা কুশিয়ারা বন্যা পাবিত অঞ্চলে যেখানে পরিবর্তন এখনো সূচিত হয়নি, সেখানে সংগৃহীত তথ্য বাস্তুসংস্থানের বর্তমান ব্যবহার এবং ওই পরিপ্রেক্ষিতে জীবনধারায় পরিবর্তনের সুষ্ঠ প্রভাব অনুধাবন করার ওপর আলোকপাত করা হয়েছে।

কেস স্টাডিতে দেখা গেছে, জীবনধারণের সুব্যবস্থাকে জলজ বাস্তুসংস্থানের সঙ্গে সম্পর্কিত করে সিদ্ধান্ত গ্রহণের জন্য অতিরিক্ত উপযোগি তথ্য তৈরি করা সম্ভব হয়েছে। ফলশ্রুতিতে এটা সুপারিশ করা যায় যে, এরকমের একটি মাপকাঠি আইডব্লিউআরএম এর যেকোন সমীক্ষার অংশ হিসেবে ব্যবহার করা সম্ভব। যদি সময় এবং সম্পদ পর্যাপ্ত পরিমাণে থাকে, তবে এই তত্ত্বে আলোচিত প্রথম মাত্রার সংখ্যাতাত্ত্বিক গণণার প্রভাব, ২য় মাত্রায় বর্ধিত করে আলোচনা করা যেতে পারে।

Bengali summary

এই তত্ত্ব মানুষের জীবনধারণের সুব্যবহার সঙ্গে পরিবেশগত পরিবর্তনের চিত্র তুলে ধরেছে, আইডবিউআরএম এর মাধ্যমে। যেখানে একটি সঠিক সিদ্ধান্ত গ্রহণের জন্য ব্যবহার করা হয়েছে, পালাক্রমে অগ্রসরমান একটি কাল্পনিক নকশা। এই পদ্ধতির মাধ্যমে তত্ত্বটি আইডবিউআরএম এবং পরিবেশগত পরিবর্তনে ভবিষ্যত কর্মপদ্ধতি সম্পর্কে একটি ভূমিকা রাখবে। সিদ্ধান্তপ্রণেতাদের জন্য তত্ত্বে বর্ণিত পদ্ধতি একটি পূর্ণাঙ্গ ধারণা দেয় এবং সামাজিক বিষয়ের সঙ্গে সংশ্লিষ্ট তথ্য প্রদান করবে, যা আইডবিউআরএম এর সামাজিক সমতার উন্নয়নে প্রয়োজনীয়।

Farsi summary
Translated by: Babak Bozorgi, Water Research Institute, Tehran, Iran.

خلاصه فارسی

ترجمه شده توسط: بابک بزرگی، مؤسسه تحقیقات آب، تهران، ایران

مقدمه

طی قرن‌ها، توسعه منابع آب با هدف افزایش سطح رفاه عمومی صورت گرفته است. در قرن گذشته، اثرات منفی ناشی از انجام فعالیت‌های توسعه منابع آب در مناطق بالادست بر روی اکوسیستم‌های مناطق پایین‌دست در سیستم رودخانه‌ها مشهود بوده است. اکوسیستم‌های رودخانه‌ها یکی از منابع تولید درآمد، غذا و سایر کالاها و محصولات برای میلیون‌ها نفر در سراسر جهان به شمار می‌روند. اگر چه توسعه منابع آب بدون شک رفاه و پیشرفت را برای تعداد کثیری از افراد به دنبال خواهد داشت، اما این منافع اغلب به طور مساوی در میان ساکنین حوضه‌های آبریز رودخانه‌ها تقسیم نشده‌اند. جوامع محلی با تأثیرات منفی ناشی از این گونه فعالیت‌های توسعه‌ای بر روی اکوسیستم‌های پایین‌دست رودخانه‌ها، در عرصه‌های ذیربط مواجه می‌شوند.

جریان‌های مورد نیاز محیط زیست به آب‌هایی که در محدوده یک رودخانه، تالاب یا منطقه‌ای ساحلی به منظور حفاظت از اکوسیستم مورد استفاده قرار می‌گیرند و منافعی که برای ساکنین این مناطق به همراه خواهند داشت، اشاره می‌نماید. روش‌های متعددی به منظور تعیین میزان جریان‌های مورد نیاز محیط زیست ارائه شده‌اند. اغلب روش‌های ارزیابی جریان‌های مورد نیاز محیط زیست بر رابطه بین جریان رودخانه‌ها و شرایط اکوسیستم تأکید داشته و به شکلی صریح اهمیت جریان‌های مورد نیاز برای تأمین رفاه گروه‌های مختلف اجتماعی را به کمیت در نمی‌آورند.

به منظور برقراری توازن در تخصیص آب در مدیریت به هم پیوسته منابع آب (IWRM)، لازم است تمامی منافع و تأثیرات منفی ناشی از تخصیص آب به هر یک از بخش‌های مصرف‌کننده آب

به کمیت درآیند. بطور سنتی تأکید بر روی منافع اقتصادی بوده است، اما امروزه پایداری زیست‌محیطی و تساوی اجتماعی نیز از جمله معیارهای مهم در ارزیابی استراتژی‌های مدیریت منابع آب محسوب می‌شوند. به منظور ارزیابی تساوی اجتماعی که به عنوان توزیع تأثیرات مثبت و منفی بر روی گروه‌های ذینفع مختلف تعریف می‌شود، باید اهمیت جریان‌های مورد نیاز محیط زیست در ایجاد رفاه برای این گروه‌ها مشخص شود.

در حال حاضر روشی به کمیت در آوردن ارزشی که جریان‌های مورد نیاز محیط زیست در تأمین رفاه افراد دارند (که از این پس به عنوان ارزش‌های جریان‌های مورد نیاز محیط زیست در تأمین رفاه جوامع انسانی نام برده خواهد شد)، وجود ندارد؛ در حالی که این روش‌ها به منظور تسهیل و ارتقاء مدیریت عادلانه ضروری هستند.

بدین ترتیب، هدف تحقیق مورد نظر در رساله حاضر عبارت است از:

توسعه دیدگاهی برای ارزیابی ارزش‌های رفاه جوامع انسانی و تساوی اجتماعی مرتبط با جریان‌های مورد نیاز محیط زیست برای کاربرد در مدیریت به هم پیوسته منابع آب

به منظور دستیابی به هدف فوق، نخست یک مدل مفهومی برای تشریح رابطه بین جریان‌های مورد نیاز محیط زیست، رفاه افراد و مدیریت به هم پیوسته منابع آب توسعه یافته و در دو مطالعه موردی مورد استفاده قرار گرفته است که عبارتند از سیلابدشت رودخانه‌های Surma-Kushiyara در کشور بنگلادش و تالاب‌های هامون در ایران.

مدل مفهومی

به منظور تشریح رفاه افراد، مؤلفه‌های مختلف رفاه مورد شناسایی قرار گرفتند. سه مؤلفه درآمد و غذا، سلامتی و آگاهی و تجربه بطور مستقیم با اکوسیستم رودخانه در ارتباط هستند. از آنجا که سیستم اقتصادی- اجتماعی، سیستمی پیچیده و پویا است، لذا تغییر در هر یک از مؤلفه‌های مذکور از تأثیر احتمالی بر سایر مؤلفه‌ها از جمله استقلال، ساختار اجتماعی و سایر عوامل روانی- اجتماعی برخوردار خواهد بود. در این رساله از این مؤلفه‌های رفاه به عنوان مؤلفه‌های درجه دوم یاد می‌شود، زیرا این مؤلفه‌ها صرفاً بطور غیر مستقیم از تغییر در اکوسیستم رودخانه‌ها تأثیر می‌پذیرند. از مؤلفه‌هایی که مستقیماً با اکوسیستم مرتبط هستند، به عنوان مؤلفه‌های درجه اول یاد می‌شود. در نتیجه ایجاد تغییر در مؤلفه‌های درجه دوم، مؤلفه‌های درجه اول نیز می‌توانند در جهت

مثبت یا منفی تغییر نمایند. رساله حاضر بر به کمیت در آوردن مؤلفه‌های درجه اول متمرکز شده است.

به کمیت در آوردن اثرات ناشی از تغییر در رژیم جریان بر روی رفاه افراد مستلزم درک مفهومی دقیقی از *روابط* بین جریان‌های رودخانه، اکوسیستم رودخانه و رفاه می‌باشد. اگرچه تأکید رساله حاضر بر رابطه بین رفاه افراد و اکوسیستم رودخانه معطوف است، اما بررسی رابطه بین اکوسیستم رودخانه و رژیم جریان نیز در قالب مدل مفهومی لازم به نظر می‌رسد. هر دو رابطه فوق به منظور ارزیابی تغییر در رفاه افراد در پی تغییر در رژیم رودخانه ضروری به نظر می‌رسند و تنها از طریق ارزیابی تمامی روابط ممکن می‌توان عدم هماهنگی بین مفاهیم و دیدگاه‌ها را شناسایی نمود.

به منظور ارزیابی اهمیت رژیم جریان و اکوسیستم رودخانه در تأمین رفاه گروه‌های مختلف اجتماعی، نه تنها بررسی رابطه بین رفاه، اکوسیستم رودخانه و رژیم جریان لازم است، بلکه مفهوم این روابط نیز باید مورد ارزیابی و بررسی قرار گیرد. منظور از مفهوم این روابط، تمامی عوامل تأثیرگذار بر رابطه بین رژیم جریان رودخانه، اکوسیستم رودخانه و رفاه است. تنها از طریق بررسی جامع بر روی این مفاهیم می‌توان اهمیت اکوسیستم رودخانه و رژیم جریان در تأمین رفاه جوامع انسانی را درک و مشخص نمود.

هدف از ارائه مدل مفهومی کمک به بررسی و تحلیل اثرات استراتژی‌های مختلف مدیریت منابع آب بر رفاه جوامع انسانی و تساوی اجتماعی است. ایجاد ارتباط بین مدل مفهومی برای گروه‌های مختلف اجتماعی که احتمالاً از تغییر در رژیم جریان رودخانه متأثر خواهند شد، حائز اهمیت است. بنابراین، شناسایی این گروه‌های مختلف اجتماعی (گروه‌های ذینفع) گامی مهم در استفاده از این مدل مفهومی است.

بکارگیری مدل مفهومی شامل پنج مرحله زیر است:

1. شناسایی گروه‌های ذینفع؛
2. ارزیابی رابطه بین رفاه جوامع انسانی و اکوسیستم رودخانه؛
3. بررسی رابطه بین اکوسیستم رودخانه و رژیم محلی جریان (ویژگی‌های هیدرولوژیکی)؛
4. ارزیابی رابطه بین رژیم محلی جریان و رژیم جریان در بالادست؛ و

5. ارزیابی اثرات ناشی از اعمال طرح‌های مدیریت منابع آب بر رفاه گروه‌های ذینفع شناسایی‌شده.

مطالعات موردی

رودخانه‌های Surma و Kushiyara بنگلادش

در رودخانه‌های Surma و Kushiyara در شمال شرقی بنگلادش، انتظار می‌رود که رژیم رودخانه در اثر سه طرح توسعه‌ای تغییر نماید:

- تغییرات مورفولوژیکی در محلی که رود Barak از کشور هند وارد بنگلادش شده و به دو شاخه Surma و Kushiyara منشعب می‌شود؛
- ساخت مخزن سد Tipaimukh بر روی رودخانه Barak (در بالادست مرز هند و بنگلادش) که به نظر می‌رسد منجر به کاهش جریان‌های حداکثر و افزایش جریان‌های حداقل شده است؛
- انحراف آب از سد Barak (در بالادست مرز هند و بنگلادش) به منظور آبیاری دشت Cachar در ترکیب با مخزن سد Tipaimukh که به نظر می‌رسد منجر به کاهش بیشتر جریان‌های حداکثر شود. افزایش جریان‌های حداقل در پی ساخت مخزن سد Tipaimukh محدود خواهد شد. مشخص نیست که تأثیر خالص بر روی جریان‌های حداقل منجر به افزایش جریان خواهد شد یا کاهش آن.

مدل مفهومی جهت ارزیابی اهمیت رژیم جریان رودخانه برای حدود ۳۰۰٬۰۰۰ نفر ساکنین سیلابدشت رودخانه‌های Surma-Kushiyara مورد استفاده قرار گرفته است. جمع‌آوری اطلاعات لازم به منظور بررسی رابطه بین رفاه افراد و اکوسیستم از طریق انجام مصاحبه در سطح خانوار انجام شد. به منظور بررسی سایر روابط از اطلاعات موجود استفاده شد.

گروه‌های ذینفع در سیلابدشت Surma-Kushiyara را می‌توان متشکل از سه دسته اصلی دانست که عبارتند از کشاورزان، ماهیگیران و سایر افراد. کشاورزان و ماهیگیران به میزان زیادی برای درآمد و غذای خود به کالاها و خدمات حاصل از رودخانه و سیلابدشت وابسته‌اند. حدود ۱۰٪ از ساکنین این منطقه از آب رودخانه برای مصارف خانگی استفاده می‌کنند. افرادی که برای مصارف خانگی آب به رودخانه وابسته‌اند، عموماً فقیرترین ساکنین منطقه می‌باشند که عبارتند از ماهیگیران

و کارگران بدون زمین. مشکلات مربوط به سیل‌گرفتگی در مناطق مسکونی عمده‌ترین رابطه بین اکوسیستم رودخانه و آگاهی و تجربه افراد به شمار می‌رود. همچنین، برای این ارزش رفاهی، افراد فقیر از بیشترین آسیب‌پذیری برخوردار هستند.

طی یک بررسی کیفی، سه نوع تغییر در رژیم جریان که در فوق به آنها اشاره شد، مورد مطالعه و بررسی قرار گرفت. نتایج این مطالعه حاکی از آن است که چنانچه دشت‌های سیلابی به موقع دچار سیل‌گرفتگی نشوند، به ویژه ماهیگیران متضرر خواهند شد. از سوی دیگر، سلامت ماهیگیران و سایر خانواده‌های فقیر نیز به واسطه افزایش جریان رودخانه‌ها در طی فصول خشک بهتر و بیشتر تأمین خواهد شد.

تالاب‌های هامون، ایران

استفاده دیگری از مدل مفهومی در قالب بخشی از یک مطالعه مدیریت به هم پیوسته منابع آب در دلتای سیستان در ایران انجام شد. دلتای سیستان در همسایگی سیستم تالاب‌های هامون قرار دارد که در شرایط مرطوب، مساحت آن معادل ۵۰۰,۰۰۰ هکتار می‌باشد. این تالاب‌ها از جهت کسب درآمد برای عده‌ای از ساکنین این منطقه و نیز جلوگیری از ایجاد طوفان‌های شن و استفاده از مناظر زیبای آن برای سایرین به منظور برگزاری جشن‌ها، از جمله جشن سال نو، از ارزش و اهمیت ویژه‌ای برخوردار می‌باشند.

به منظور بررسی اهمیت و کاربرد این مدل در ارزیابی کمی رفاه افراد در شرایط استراتژی‌های مختلف مدیریت منابع آب، پنج مرحله مختلف مدل مفهومی انجام گرفت. به منظور جمع‌آوری اطلاعات در مورد رابطه بین رفاه افراد و اکوسیستم تالاب‌ها، مباحثات گروهی و تحقیق از طریق تکمیل پرسشنامه انجام شد. به منظور بررسی سایر روابط موجود نیز مطالعات انجام شده در قالب مطالعات مدیریت به هم پیوسته منابع آب بکار گرفته شد.

با در نظر گرفتن وابستگی ساکنین منطقه به تالاب‌های هامون، جمعیت دلتای سیستان را می‌توان متشکل از سه گروه مختلف دانست. از تعداد تقریبی ۷۱,۰۰۰ خانوار این منطقه، حدود ۱۵,۰۰۰ خانوار برای بیش از 70 درصد درآمد خود متکی به تالاب‌های هامون بوده، حدود ۳۰,۰۰۰ خانوار برای بیش از 70 درصد درآمد خود به کشت آبی متکی هستند و درآمد ۲۶,۰۰۰ خانوار دیگر که در مناطق شهری ساکن می‌باشند به طور مستقیم با تالاب‌ها مرتبط نیست.

عمده‌ترین محصولات و خدماتی که تالاب‌ها برای ساکنین منطقه فراهم می‌آورند عبارتند از ماهیان، پرندگان و نیزارها که درآمد و غذای این افراد را تأمین نموده، همچنین مانع از بروز طوفان‌های شن شده و در تنظیم و تعدیل آب و هوای منطقه نیز مؤثر می‌باشند. پنج عامل هیدرولوژیکی مهم در این تالاب‌ها شناسایی شدند که در بقای این کالاها و خدمات مؤثر هستند. سرریز شدن آب تالاب‌ها، خشکسالی‌های منظم، یک مقدار حداقل از سطح آبگرفتگی در فصل پاییز و یک مقدار حداقل حجم جریان ورودی در فصل بهار از جمله عواملی هستند که باعث حفظ حیات ماهیان، نیزارها و پرندگان در این تالاب‌ها می‌شوند. همچنین، یک مقدار حداقل آبگرفتگی در تالاب هامون صابری از ماه می تا آگوست (اردیبهشت تا مرداد) به منظور جلوگیری از بروز طوفان‌های شن از جمله مسائل دیگر حائز اهمیت می‌باشد.

استراتژی‌های مدیریت منابع آب از طریق اثرات خود بر رژیم هیدرولوژیکی تالاب‌ها، تغییرات متعاقب آن در اکوسیستم و فراهم شدن کالا و خدمات اکوسیستم سطح رفاه گروه‌های ذینفع مختلف را تحت‌الشعاع قرار می‌دهند. دیدگاهی برای به کمیت در آوردن تغییرات ایجادشده در رفاه گروه‌های ذینفع در پی تغییر در اکوسیستم تالاب‌ها در رابطه با تمام عوامل تأثیرگذار بر رفاه مورد بحث قرار گرفته است. از طریق این روابط، اثرات وارد بر گروه‌های مختلف ذینفع و تساوی در توزیع این اثرات وارده مورد ارزیابی قرار گرفت.

عمده‌ترین بخش‌های تقاضای آب، کشت آبی و تالاب‌های هامون می‌باشند. هر دو این بخش‌های مصرف‌کننده آب برای اقتصاد منطقه و رفاه گروه‌های ذینفع حائز اهمیت هستند. لازم است توازنی برای توزیع آب بین این دو بخش برقرار شود. به منظور شناسایی چنین توازنی، استراتژی‌هایی با تأکید بر افزایش و کاهش سطح زیر کشت آبی مورد تحلیل قرار گرفتند. نتیجه حاصل این است که در شرایط حاضر، توزیع اثرات از بیشترین توازن برخوردار است. اگر چه تحت شرایط فعلی و شرایط کاهش سطح زیر کشت آبی، شرایط مشابهی برای اکوسیستم محاسبه شد، اما تعداد سال‌های همراه با کاهش درآمد برای گروه‌های ذینفع مختلف تغییر نمود. این مطلب حاکی از آن است که ارزیابی اثرات ناشی از رفاه افراد اطلاعات مفیدی را فراهم خواهد نمود که به نوبه خود منجر به ارتقاء سطح مدیریت آگاهانه و عادلانه خواهد شد.

نتیجه‌گیری

از طریق شرح روابط بین رفاه افراد، اکوسیستم رودخانه و رژیم جریان و ماهیت این روابط، مدل مفهومی پیشنهادی، به کمیت در آوردن تغییرات ایجادشده در رژیم جریان یا اکوسیستم رودخانه بر

اساس رفاه افراد و تساوی اجتماعی را به عنوان بخشی از تحلیل‌های مدیریت به هم پیوسته منابع آب تسهیل می‌نماید.

در حالی که مدل مفهومی و دیدگاه مرحله‌ای را می‌توان بطور کلی مورد استفاده قرار داد، در عین حال لازم است روش‌های جمع‌آوری اطلاعات و تحلیل‌ها متناسب با شرایط حاکم ارائه شوند. در منطقه خشک سیستان، مشخص‌نمودن محدوده اکوسیستم تالاب‌ها و شناسایی افراد ذینفع از این تالاب‌ها به راحتی قابل انجام است. در منطقه مرطوب شمال شرقی بنگلادش، تعیین میزان وابستگی محصولات و خدمات اکوسیستم به جریان رودخانه‌ها دشوارتر به نظر می‌رسد و لذا تعیین این که آیا مصرف‌کنندگان این کالاها و خدمات ذینفعان متأثر از رژیم رودخانه هستند یا خیر، به مراتب دشوارتر می‌باشد. بعلاوه، این حقیقت که پیش از این مداخلاتی در سیستم رودخانه در منطقه سیستان صورت گرفته، امکان انجام گفتگوهای مستقیم در رابطه با اثرات تغییر در تالاب‌ها و نقش آن در تأمین رفاه ذینفعان مختلف را با آنان فراهم می‌آورد. در سیلابدشت Surma-Kushiyara که تا کنون تغییراتی در آن صورت نگرفته، جمع‌آوری اطلاعات بر نحوه استفاده فعلی از اکوسیستم و شناسایی شرایط موجود متمرکز است که با هدف پیش‌بینی اثرات تغییر در رژیم جریان بر رفاه افراد صورت می‌پذیرد.

طی مطالعات موردی، این نتیجه حاصل شد که از طریق ارزیابی نقش اکوسیستم رودخانه‌ها در رفاه افراد، اطلاعات مفید دیگری نیز به دست آمد که به تصمیم‌گیری‌ها کمک می‌کند. بنابراین پیشنهاد می‌شود که این‌گونه ارزیابی‌ها در قالب مطالعات مدیریت به هم پیوسته منابع آب صورت گیرند. چنانچه زمان و منابع محدود باشند، ارزیابی رفاه افراد را می‌توان تنها به صورت نوعی ارزیابی انجام داد که تنها بر شناسایی گروه‌های ذینفع و اهمیت کالاها و خدمات مختلف اکوسیستم برای هر یک از این گروه‌های ذینفع تأکید خواهد داشت. چنانچه زمان و منابع کافی در اختیار باشد، در آن صورت تعیین اثرات درجه اول که در این رساله به شرح آنها پرداخته شده را می‌توان با تعیین اثرات درجه دوم توسعه داد.

در این رساله، دیدگاهی به منظور ارزیابی نقش ارزش‌های رفاه اجتماعی جریان‌های مورد نیاز محیط زیست به عنوان بخشی از مطالعات مدیریت به هم پیوسته منابع آب ارائه شده است که متشکل از یک مدل مفهومی و نیز دیدگاهی مرحله‌ای به منظور ارزیابی واقعی می‌باشد. این رساله به واسطه ارائه دیدگاه مذکور در عملیاتی نمودن هر چه بیشتر مدیریت به هم پیوسته منابع آب و جریان‌های مورد نیاز محیط زیست مؤثر واقع خواهد شد. دیدگاه ارائه‌شده در این رساله اطلاعات

جامع‌تر و مرتبط با اجتماع را برای تصمیم‌گیرندگان فراهم می‌آورد که در ارتقاء هر چه بیشتر تساوی اجتماعی در مدیریت به هم پیوسته منابع آب ضروری می‌باشد.

Acknowledgements

This PhD thesis would not have been the same without the support and input from various people. Eelco van Beek, thank you first of all for the opportunity to let me do this research and start work at Delft Hydraulics at the same time. I am grateful for your guidance in the process of PhD research which kept me moving forward. Thanks to Frans Klijn en Rinus Vis for providing detailed comments on what I had written in various stages.

This research gave me the opportunity to work both in Bangladesh and Iran. It has been an amazing experience to travel and meet the people of these cultures, both at the research institutes in the capital and in the villages in the field.

Many thanks go to Asha Naznin, my research assistant in Bangladesh, who also made the Bengali translation of the summary. Without you I don't know what I would have done in the villages of Northeast Bangladesh. I am impressed by your way of thinking and living in this complicated country and hope you continue your work on gender issues as a journalist! Thanks also to Prof. M.F. Bari, Mohammed Mukturuzamman, Marcel Marchand and Bas de Jong for helping me with my research in Bangladesh in various ways.

My work in Iran greatly benefited from the help of Sara Hajiamiri. Thanks also to Babak Bozorgi for translating the summary into Farsi. The interaction with colleagues at the Water Research Institute made missions fun and interesting. I have come to appreciate Iranians as polite and kind people with a rich culture.

Thanks to Louise Korsgaard, Michael Moore and Katharine Cross. Our meetings have always been so much fun and have strongly motivated me. I am so happy we have met and hope we will continue work on the Environmental Flows Network, together with people from the various institutes involved.

Thanks to all members of the promotion committee – Ainun Nishat, Reza Ardakanian, Guus Borger, Arjen Hoekstra, Huub Savenije and Huib de Vriend – for taking the time to read and comment on my thesis and for travelling to Delft to take part in the defence ceremony. I owe special thanks to Jackie King, the discussions we have had the few times we have met have been extremely inspiring and have given direction and shape to my thesis. I am happy that you have also commented on and approved the final thesis.

Thanks to many colleagues at both Delft University of Technology and Delft Hydraulics. The many coffee breaks and drinks after work were good to remember that there is life beyond PhD research.

Acknowledgements

Thanks to all family and friends for understanding that often I didn't have time. I really hope we will be able to see more of each other from now on! And lastly I like to thank my 'own' Eelco for being there for me. We are having a fantastic summer ahead of us!

Curriculum Vitae

Karen Meijer was born in Assen, the Netherlands, on August 3rd, 1976. She attended high school at the Christelijk Lyceum Apeldoorn and received her VWO diploma in 1994. After one year of environmental studies at Wageningen Agricultural University, she started studying Civil Engineering and Management at the University of Twente in 1995. In December 2000 she received her masters degree. She wrote her master thesis on the impacts of concrete lining of irrigation canals on the availability of water for local communities in Uda Walawe, Sri Lanka, at the International Water Management Institute (IWMI) in Sri Lanka.

In August 2001, Karen started her PhD research on the quantification of well-being values of environmental flows for IWRM in a part-time position at Delft University of Technology, department of Water Management. At the same time she started working part-time at WL | Delft Hydraulics. Together with researchers from other institutes she was involved in the initiation of the Global Environmental Flows Network. Since January 2006, Karen is a full-time employee of the River Basin Management group of WL | Delft Hydraulics.